Consequences of Party Reform

CONSEQUENCES
OF PARTY REFORM

NELSON W. POLSBY

OXFORD UNIVERSITY PRESS
Oxford   New York   Toronto   Melbourne
1983

Oxford University Press
Oxford London Glasgow
New York Toronto Melbourne Auckland
Delhi Bombay Calcutta Madras Karachi
Kuala Lumpur Singapore Hong Kong Tokyo
Nairobi Dar es Salaam Cape Town

and associate companies in
Beirut Berlin Ibadan Mexico City Nicosia

Library of Congress Cataloging in Publication Data
Polsby, Nelson W.
Consequences of party reform.
Includes index.
1. Presidents—United States—Nomination. I. Title.
JK521.P58 1983 324.5'0973 82-14509
ISBN 0-19-503234-9
ISBN 0-19-503315-9 (pbk.)

Printing (last digit): 9 8 7 6

Printed in the United States of America

The Democratic party of the nation ain't dead, though it's been givin'
a lifelike imitation of a corpse for several years. . . . The trouble is
that the party's been chasin' after theories and stayin' up nights readin'
books instead of studyin' human nature and actin' accordin' . . . .

. . . Have you ever thought what would become of the country if
the bosses were put out of business, and their places were taken by a
lot of cart-tail orators and college graduates? It would mean chaos.

William L. Riordan, *Plunkitt of Tammany Hall* (1905) (New York: Dutton, 1963),
pp. 88, 81.

It has become fashionable among a certain group of pundits and political scientists to blame the reforms undertaken by the Democratic
Party after its 1968 convention for all manner of political ills—declin-
ing voter participation, the proliferation of single-issue groups, weak
presidential leadership, poor congressional performance, and, above
all, the decay of the political party. So great has been the hostility to
these reforms, to the reformers who perpetrated them and to the proliferation of presidential nominating primaries that followed in their
wake, that one expects any day now to see a book published that blames
these reforms, not only for our political ills, but for cancer, heart disease, and falling arches.

Curtis Gans, "How the White House is Won," *Book World, Washington Post* (August 12, 1979), p.10.

The processes of government are essentially educational processes.

Felix Frankfurter, letter to Franklin D. Roosevelt, November 8, 1928. In Joseph P. Lash, *From the Diaries of Felix Frankfurter* (New York: Norton, 1979),
p.40.

For
Allen I. Polsby
and
Daniel D. Polsby

Contents

Preface

This book comes to its readers through the courtesy of Watergate. For many Americans, Watergate was a profoundly unsettling business, including, as it did, the unprecedented resignation of a Vice President under a cloud of criminal charges, the near impeachment and resignation of a President, and numerous other misfortunes. Some of us who study American politics felt during those difficult times that it was a part of our professional obligation to stop whatever we were doing and pay attention to Watergate so that we could make an effort at interpreting for our fellow citizens—perhaps only slightly more confused than we were—what was going on in the light of our understanding of the ongoing patterns and requirements of the American constitutional order. For a few of us, the experiences of those years turned out to be more than a temporary detour; they more or less permanently changed our intellectual agendas. I count myself as one of those who as a result of the concerns of the Watergate era have undertaken to attend on a more or less regular basis to the state of the American political system as a whole. I thus have become interested in monitoring changes in the vital signs of the system, and—no matter how unreasonably ambitious the effort has seemed—in tracing out how changes in one part of the system affect changes in others.

As it happens, the last twenty years have been rich in these sorts of systemic changes,[1] and the purpose of this book will be to follow one set that has caused particular concern: the reform of the presidential nomination process. Because the consequences of reform in this sector have rippled out very widely into the political system, it would have been impossible to render a full account of my views within the pages of the text on presidential elections that Aaron Wildavsky and I have written together in happy intermittent collaboration since 1964.[2] It gives me great pleasure to acknowledge the constructive influence that my friend and colleague has had on this offshoot of our common enterprise.

I owe to a number of former students a debt of gratitude for the intellectual stimulation they have given me as I have wrestled with the issues of this book. Many of them are now well launched on promising careers of their own that have spread more widely the insights that once circulated around the seminar table in Berkeley. As will be readily apparent from the footnotes, I have frequently drawn from their published work, but in addition I wish to mention here James I. Lengle, Elaine Kamarck, Robert T. Nakamura, Bill Cavala, and the late Jeffrey Pressman. In a category all his own is Byron Shafer, whose prize-winning doctoral dissertation on the reforms of the Democratic party[3] provided many occasions of enlightenment, and whose energy, generosity, and high intellectual standards have contributed a great deal to these pages.

A year in England (in 1977–78) and a month in Australia and New Zealand (June 1980)—the former under John Simon Guggenheim Foundation and the latter under USICA auspices—convinced me that the proper workings of the American political system are not exclusively the concern of Americans. In the course of fifty or so talks to foreign audiences, and in response to their comments and queries, I shaped much of the argument of this book. Notable among those friends who have shared with me their valuable comparative outlooks on American politics are Leslie Stone of the BBC overseas service, Greg Armstrong at Melbourne, John Hart at Australian National University, Louis Heren, formerly of *The Times* of London, the late Robert McKenzie of the London School of Economics, Jim Sharpe and Philip Williams at Nuffield College, Oxford, Anthony King and Graham Wilson of Essex, Esmond Wright in London, Colin Seymour-Ure and Alec Barbrook at Kent, Malcolm Shaw of Exeter, Richard Hodder-Williams of Bristol, and those two pillars of American political studies in the United Kingdom, John Lees of Keele and David Morgan of Liverpool. Douglas Wilson and, especially, the incomparable Allan Croghan, sometime cultural affairs officers at the U.S. Embassy in London, were extraordinarily helpful in providing the introductions that mean so much in a society where one needs to be introduced. Geoffrey Smith of *The Times* of London has over the last few years been the foreign observer of American politics whose questions and observations have stimulated me most. Two great cosmopolitan spirits, Constance and William Drower, have also taught me much about America, in part from the fund of knowledge they accumulated in the decade Bill was the British embassy's brilliantly successful specialist on Capitol Hill in Washington.

Many of the ideas expressed in these pages have found preliminary expression in essays I have written on earlier occasions.[4] I doubt I should ever have pulled together the entire argument that was floating around in bits and pieces in my head if I had not been invited to try my hand first at one, then at another part of the overall problem. And so I am grateful to William Lee Miller, Robert Goldwin, Seymour Martin Lipset, James Sterling Young, and Walter Dean Burnham and Martha Wagner Weinberg for chances to do some thinking out loud.

I am also very grateful for three opportunities to meet with students of various aspects of reform of the presidential nomination process under the auspices of the American Bar Association: in Tiburon, California, in April 1975; in Washington

D.C., in February 1976; and at the Johnson Foundation's conference center, Wingspread, in July 1981. I likewise acknowledge the hospitality of the American Enterprise Institute and Kenyon College for inviting me to participate in a most illuminating conference at Kenyon in April 1979, and of James Sterling Young for his generous invitations to a series of Miller Center meetings on the presidency and the party system at the University of Virginia in 1979–80. Among my mentors on one or more of these occasions were Joel Fleishman, Herbert Alexander, George Agree, William Crotty, Roger Allan Moore, John Sears, William Brock, John Hoving, Stephen Gottlieb, Richard Scammon, Robert Horwitz, Tom Mann, James Ceaser, Michael Robinson, Richard Cheney, Wayne Grandquist, Norman Ornstein, David B. Truman, and Richard E. Neustadt. I emphasize that none of these distinguished people necessarily agrees with any part of my analysis, but that I found their comments thoughtful, stimulating, reassuring, informative, or all of the above. In some cases we spent hours together in informal dialogue, the infallible sign of a well-run conference.

Closer to home, I am pleased to acknowledge the intellectual influence of Allan P. Sindler, Austin Ranney, Raymond E. Wolfinger, John Zaller, Eric Davis, Herbert McClosky, Paul Sniderman, Richard Brody, and Charles O. Jones on my understanding of many of the problems discussed in this book.

The following friends and colleagues read a draft of this manuscript and offered comments, criticisms and numerous improvements: Aaron Wildavsky, Daniel D. Polsby, Larry Bartels, John Zaller, Leon Epstein, Herbert Alexander, Adam Clymer, Roger Allan Moore, John Bibby, James I. Lengle, David B. Truman, Eric L. Davis, Joel Fleishman, Douglas Bennet, James Ceaser, David Cohen, Richard Stearns, William J. Crotty, Charles O. Jones, Byron Shafer, Stephen Snyder, David Price, Patricia Brown, Morley Winograd, Donald Pfarrer, Christopher Arterton, and Everett Carll Ladd. In addition, Anne Wexler very kindly and promptly looked over a portion of the manuscript in which her name appears, and Edwin Epstein and

Pamela Goldschmidt helped dig up some hard to find data. As the length of this list should indicate, I have been the recipient of so much good advice and encouragement that it is without doubt my pig-headedness alone that accounts for the mistaken ideas in the pages that follow.

I have had the benefit of cheerful and highly competent research assistance in bringing this book to completion from Deborah Cichon and Lynne Gordon, and in addition collegial support and entrepreneurial energy of a special kind from Michael Goldstein and Peverill Squire. The Department of Political Science, the Committee on Research, the Survey Research Center, the Institute of Governmental Studies and its National Policy Studies Program, the Graduate Division and the Office of the Chancellor, all of the University of California, Berkeley, have in one way or another contributed to the ongoing enterprise that made this book possible. In this connection I would like to thank Sanford S. Elberg, William A. Shack, William K. Muir, Eugene C. Lee, and Percy Tannenbaum, each of whom gave at the office. In all these thank-yous, a reader will be unable to find an extraordinary extra-mural grant of funds. This book was produced without one, a tribute to the resourcefulness of the University of California at Berkeley, and to its commitment to routinely providing assistance, and the right sort of atmosphere, for scholarship.

The two learned attorneys to whom this book is dedicated have not only given me shelter, support, laughter, and good counsel for most of my life, but have done so in a spirit for which the term brotherhood is, as it happens, an excellent approximation.

Finally, I thank Daniel R. Polsby, Emily A. Polsby, and Lisa S. Polsby for being unusually bountiful with their affections and sparing with their disaffections all these years. My most skilled editor and closest friend Linda O. Polsby shares with me the responsibility for the three last named, but wishes to be excused along with everyone else from responsibility for the various mistakes and lapses in judgment that not even her unblinking eye, sharp pencil, and impeccable judgment could prevent in the pages that follow.

Like many others who are concerned with political parties, presidential selection, and reform, I am looking forward to reading the results of the many projects, study groups, and commissions that have lately been formed and funded to consider these and related matters. As an early offering in the field, this book can hope to do a little agenda-setting for those that follow, and contribute to ongoing discussion as much through the questions it has formulated but answered inadequately as through those it has settled. At least it would be nice to think so.

Berkeley, California   N. W. P.
March 15, 1982

Consequences of Party Reform

Introduction

It is not at all unusual for Americans to find fault with the alternatives available to them on any given election day. No presidential election within living memory has been totally devoid of dissatisfactions focused on the shortcomings, real or imagined, of the main candidates and of the processes that brought them forward.[1] Yet it may not be wholly an optical illusion stimulated by the proximity in time of the last few elections that suggests that the contemporary presidential selection process does pose serious problems for the American political system. Consider one straw in the wind: by the simple criterion of prior experience in national politics and government no newly elected President in the entire history of the republic was ever less well prepared to take office than Jimmy Carter, unless it was Ronald Reagan.[2] That our two most recent Presidents should be arguably the two most nationally inexperienced in American history cuts directly across the grain of common sense that tells us that the duties of the Presidency are not smaller today than when more seasoned candidates—whatever their other qualifications and limitations—were routinely the only ones considered suitable for the job.

A criterion such as prior experience is an elite, if not an elitist, gauge with which to measure a presidential candidate in that it is likely to be invoked primarily by those who are well enough informed about politics to have some notion of the on-the-job benefits that a fund of exact information may confer on a newly elected President. This may suggest the extent to which the American system of presidential selection has moved away from elite and toward mass criteria in providing means for assessing the qualifications of presidential candidates.[3] To observers accustomed to democratic political systems that routinely choose their heads of government only from a pool of those previously elected to the national parliament and then selected by colleagues to lead a parliamentary party, the inattention of the American presidential selection process to the issue of prior preparation seems unfamiliar, uncomfortable, and even dangerous.[4] Some of them are genuinely puzzled that the American political system has in recent years appeared to move, if in any direction, away from concern with the political education, the intellectual capital, of what many of them conceive to be the most important leader in the world.

Even more interesting is the possibility that such a trend might be accounted for in part by a series of explicit choices made by politicians. The purpose of this book is to explore this possibility. It seeks to establish a few connections between causes and effects in the loosely coupled, overdetermined, busy world of American politics, and attempts to trace through some of the effects of some political reforms, choosing from a wider menu of possibilities those recent reforms that seem to have had an impact on later events sufficiently important to warrant thoughtful reconsideration. It is not the argument of this book, however, that the explicit choices entailed in the political reforms presently to be considered are completely responsible for the consequences I may seem to be fastening on them. I wish rather to argue that they are partially responsible, and facilitative in character, that they were necessary but not sufficient pre-conditions of a particular set of institutional problems toward which the American political system evolved in the wake

of the turmoil of the late 1960's. Although other forces of varying sorts were also at work—technological, demographic and
institutional changes that were bound, for example, to increase
the independent influence of the news media on American politics—it does not seem to me unreasonable to focus on the contribution to social change made by the explicit choices of political actors. To study political choices in light of their
consequences is not to claim they are the only determinants of
consequences but merely to acknowledge that they are a set of
determinants through which people took a direct hand in man-
aging their own affairs, and might have managed differently,
and that studying how these choices are made, their factual
premises, their results, and their theoretical justifications can
hope to add to usable knowledge.

Given the rumbles of dissatisfaction in both major parties
about the political arrangements resulting from the last wave
of reform, the practical implications of a study of this sort seem
clear enough. In addition, this study may also contribute in at
least three ways to political science: as an opportunity explicitly
to consider interrelations among institutions of the American
political system normally studied in isolation from one another;
as an exploration of the process of reform and of ways to manage it so as to minimize the burden of unexpected and unwanted consequences; and as an example of the sort of policy
analysis that falls squarely within a subject matter area where
political scientists carrying a brief primarily for the understanding of process might supplement and criticize recommendations of actors more concerned with the manipulation of outcomes for their own partisan or intramural advantage.

The argument to be made in this book may or may not prove
to be appealing when its underlying assumptions about political
life and human behavior are taken into account: that changing
the rules of politics changes the incentives for political actors;
that changing incentives leads to changes in political behavior;
and that changing behavior changes political institutions and
their significance in politics. Thus we shall be exploring the

educational impact of political reform, the capacity of rules of
conduct actually to guide and to influence the activities of a
varied cast of political actors: candidates for public office and
their managers, state and local party leaders, interest group
leaders, delegates to national party conventions, journalists and
television news producers, and ordinary voters.

The discussion begins with a short historical account of the
reasons for the adoption of the two main sets of party reforms
whose interaction together and with broader trends has produced significant consequences, namely, reforms of the delegate selection process and reforms of party finance.[5] The second chapter outlines major consequences of these reforms
within their intended sphere of influence, the political parties,
and shows how the system of party competition as well as the
individual party organizations have been affected by reform. Because parties and the party system have been changed, the chief
results of party activity at the national level—recruitment of
presidential candidates—have in turn had an impact upon the
conduct of the presidency. This, at any rate, is the argument
of Chapter Three. Chapter Four examines wider consequences
of reforms in the political system, for interest groups, for the
news media, and for processes of political intermediation more
generally, and for the political mobilization and participation
of ordinary citizens. The final chapter returns to the general
subject of reforms and, in the light of the consequences considered in previous chapters, reformulates the criteria by which
changes in the party system might be justified and evaluated.

It is to be expected that there might well be reasonable disagreement with many of the conclusions reached in the course
of this inquiry. Even if this proves to be the case, it is desirable
to narrow the ground upon which policy disagreement might
rest to differences of opinion rather than differences in empirical knowledge. Thus at every feasible juncture I have taken
pains to establish the factual premises of my argument with
such empirical information as I could gather, and to indicate
those instances where the empirical information I could bring

to bear on a given question seemed to me to fall short of being satisfactory. On the whole, I am content to say that the line of argument taken in this book does no great violence to the facts as best I presently can marshal them, but one must concede that in a world as tumultuous and diverse as the world of American politics, there are quite likely more than enough facts available to lend themselves to arrangement in still other patterns, some of which may flatly contradict some of the conclusions of this book. Indeed, one of the scientific purposes of a book of this sort is to so sharpen the perceptions of its readers as to make more visible the bases for rejecting, modifying, and reformulating, not merely for accepting its central propositions. This invitation to disagreement is, perhaps, the chief distinction which ordinarily separates a work of political science from a work of political persuasion, and should, unlike the case of a work of political persuasion, make it more persuasive.

I

The Party Reforms
and How They Grew

1. Eyewash? Primaries Before Reform

Perhaps the most famous remark ever made about American presidential primary elections was Harry S Truman's declaration on January 31, 1952, that they were "eyewash." The circumstances under which he said it were suspect: Senator Estes Kefauver of Tennessee, whom he disliked, was running hard and essentially unopposed in the New Hampshire Democratic primary, then as now the first of the election season.[1] Moreover, recent memory suggested that primaries could not be quite so lightly dismissed. In 1944 Wendell Willkie's presidential candidacy had been done in by the Wisconsin primary just four years after he had won the Republican nomination on the eighth ballot at the National Convention of 1940.[2] In 1948 Harold Stassen had risked a radio debate with Thomas E. Dewey as a feature of his campaign in the Oregon primary. He lost the debate, the primary, and the nomination.[3]

Later on in 1952—on March 18—Dwight D. Eisenhower received 100,000 write-in votes in the Minnesota Republican primary. Though it was less than the vote accorded favorite son Harold Stassen, it overshadowed Eisenhower's New Hampshire

victory of the previous week and launched his successful campaign for the nomination of his party.[4]

On the Democratic side, however, primaries that year did
prove to be eyewash: Kefauver won most of them but lost the
nomination.[5] He entered the Democratic National Convention
with 257½ delegates pledged to his candidacy, many more than
the 161½ pledged to Senator Richard Russell of Georgia, the
112½ pledged to New York's Averell Harriman, the 45½ to
Senator Bob Kerr of Oklahoma, or the 41½ to Governor Adlai
Stevenson of Illinois. His total still fell far short of the 616
needed to win the nomination. As the convention opened, 611½
votes, nearly a majority, were spread among minor candidates,
were formally uncommitted, or were in dispute.[6] It was the last
major party convention in American history—there have been
fourteen conventions since—to have taken more than a single
ballot to nominate a presidential candidate.

On the first ballot Kefauver got 340 votes, more than any of
the other thirteen candidates for whom votes were recorded.
On the second ballot he peaked at 362½ votes, and on the third
he slipped back to 275½ votes, as Adlai Stevenson gathered in
just enough of the scattered votes to win with 617½.[7] There
was nothing unexpected about Stevenson's eventual win. He had
been the choice of the incumbent President of his own party,[8]
acceptable to labor, as another attractive centrist candidate, Vice
President Alben Barkley, was not,[9] moderate enough to be un-
threatening to the South—as his Illinois colleagues had under-
scored by voting earlier in the convention to seat challenged
delegates from Virginia[10]—and well connected with the Eastern establishment—as Averell Harriman's well-timed withdrawal in his favor attested.[11] Indeed, with so many of the forces
in the Democratic party moving in his direction, it was a wonder that his nomination took so long. The delay can be accounted for mainly by Stevenson's obdurate reluctance to seem
to be doing anything at all in his own behalf.[12] This momentarily stalled his bandwagon and exasperated many of his supporters, but in the end did not prevent the Democratic party
from making him the nominee.

Stevenson had entered no primaries, and initially won no delegates pledged by primary elections. He beat Kefauver, the winner of 174 delegates in this fashion, and the leader in number of committed delegates until the convention's third ballot. In 1952, though primaries counted for enough to damage a candidate's chances or dramatically boost them, winning even a large proportion of them was not a sufficient condition for winning the nomination. Indeed, from the time of their introduction into American politics around the turn of the twentieth century, until well into the 1960s, primary elections were never the sole determinant of party presidential nominations.[13]

The fact that Kefauver was a serious candidate at all was not merely a tribute to his indefatigable campaigning. He was the

Table 1.1   Winner of most primary elections: 1912–1980

Year	Republicans	Democrats
1912	Theodore Roosevelt	*Woodrow Wilson
1916	Albert Cummins	*Woodrow Wilson (incumbent)
1920	Hiram Johnson	William McAdoo
1924	*Calvin Coolidge	William McAdoo
1928	*Herbert Hoover	*Alfred E. Smith
1932	Joseph France	*Franklin D. Roosevelt
1936	William Borah	*Franklin D. Roosevelt (incumbent)
1940	Thomas Dewey	*Franklin D. Roosevelt (incumbent)
1944	*Thomas Dewey	*Franklin D. Roosevelt (incumbent)
1948	Harold Stassen	*Harry S Truman (incumbent)
1952	Robert Taft	Estes Kefauver
1956	*Dwight Eisenhower (incumbent)	Estes Kefauver
1960	*Richard Nixon	*John Kennedy
1964	*Barry Goldwater	*Lyndon Johnson (incumbent)
1968	*Richard Nixon	Eugene McCarthy
1972	*Richard Nixon (incumbent)	*George McGovern
1976	*Gerald Ford (incumbent)	*Jimmy Carter
1980	*Ronald Reagan	*Jimmy Carter (incumbent)

*Party nominee

Sources: Robert A. Diamond (ed.), *Guide to U.S. Elections* (Washington, D.C.: Congressional Quarterly Inc., 1955), pp. 58–95, 309–35; Diamond (ed.), *Presidential Elections Since 1789* (Washington, D.C.: Congressional Quarterly Inc., 1975), pp. 110–14, 119–55; Paul T. David, Ralph M. Goldman, and Richard C. Bain, *The Politics of National Party Conventions* (Washington, D.C.: Brookings Institution, 1960), pp. 213–44; *Congressional Quarterly Weekly Reports* (July 10, 1976), pp. 1794–99, (August 14, 1976), pp. 2188–96, (July 12, 1980), pp. 1928–37, (August 9, 1980), pp. 2268–76.

first national politician in American history to attain prominence by means of television. As chairman of a special Senate
committee investigating organized crime, he had earned the
enmity of "regular" Democratic politicians well beyond the states
of Illinois and Florida, where early committee hearings embarrassing to the party had been held. It was widely believed that
his committee's activities had damaged the re-election chances
in 1950 of the Senate Democratic leader, Scott Lucas of Illinois.[14] Something of a loner in the Senate, considered a rene-
gade by many of his less liberal southern colleagues, Kefauver
was not popular with Democratic officeholders—President
Truman, in particular.

But his popularity was high with television viewers after the
broadcast of the New York hearings of the Kefauver Committee, the week of March 12, 1951.[15] As Kefauver's biographer
said:

> Overnight, millions had become familiar with the tall, lanky forty-
> seven-year-old Tennessean, whose soft Southern accent exhibited
> just a trace of Appalachian influence and whose firm but fair di-
> rection of the hearings evoked almost universally favorable com-
> ment. The dignity and easily shocked innocence of Kefauver and
> the other committee members allowed millions of viewers to iden-
> tify with them . . . .[16]

> The New York investigation became the most discussed topic in
> the country, and to millions the crime committee members be-
> came overnight celebrities . . . .[17]

> The twelve months preceding the Kefauver Committee's hear-
> ing in New York City had seen the percentage of homes in the
> New York metropolitan area with TV sets rise sharply, from 29
> to 51 per cent. . . . In the morning hours, the hearings created
> seventeen times the normal viewing audience (26.2 per cent of
> homes vs. the normal 1.5 per cent); in the afternoon there was
> also a dramatic increase in viewing (from 11.6 per cent of homes
> to 31.5 per cent). It was estimated that an average of 86.2 per
> cent of those viewing television watched the hearings, and that an
> average of 69.7 per cent of the TV sets in the New York area
> were on during the hearings, twice as many as during a weekday
> World Series game in October 1950.[18]

[The hearings] were eventually carried by TV stations in twenty cities along the eastern seaboard and in the Midwest . . . .[19]

Life magazine commented:

The U.S. and the world had never experienced anything like it . . . All along the television cable . . . [people] had suddenly gone indoors—into living rooms, taverns and clubrooms, auditoriums and back offices. There, in eerie half-light, looking at millions of small frosty screens, people sat as if charmed. For days on end and into the nights they watched with complete absorption . . . the first big television broadcast of an affair of their government, the broadcast from which all future uses of television in public affairs must date . . . Never before had the attention of the nation been riveted so completely on a single matter. The Senate investigation into interstate crime was almost the sole subject of national conversation.[20]

Kefauver's biographer continues:

Newspapers were full of stories of neglected housework, and deserted movie theaters and department stores; in New York, Consolidated Edison had to add a generator to supply power for all the TV sets being used . . . .[21]

The televising of the New York hearings . . . transformed Kefauver into a genuine national hero and vastly heightened his availability for the Presidency.[22]

The year 1952 marks the beginning of the interaction between television popularity and success in the primaries. Because primary elections did not select most delegates to the Democratic National Convention in that year, criteria other than television popularity retained their importance and determined the outcome of the nomination process.

Even in 1952, however, primaries were not negligible. At a minimum they provided information to the political leaders in whose gift the presidential nomination still remained: information about the popularity of various candidates in assorted states of the Union and among admittedly restricted but indubitably mass electorates in trial-heat situations, and information also about candidates' abilities to conduct the sort of arduous face-to-face campaigning that the primary contests traditionally demand. Performance in primaries would communicate something about a candidate's stamina, perhaps about the strength of his desire to capture the Presidency, his capacity to mobilize resources in his own behalf, to meet the press early in the morning and late at night without committing some fatal error, to attract support from local party organizations, to find the right grounds in differentiating himself from competitors, yet at the same time appealing to voters. National party leaders could draw contrary inferences from data of this sort if they chose to do so. But often enough some measure of consensus arose: that Willkie had alienated too many regular Republicans by 1944, for example, that Stassen had blundered in his handling of the Oregon debate of 1948, that Eisenhower's popularity at the grass roots was overwhelming in 1952.[23] Moreover, primary elections were not all beauty contests, separated from delegate selection procedures. Some delegates were selected in primaries which might, given a close convention, make a difference in the outcome. A resourceful candidate, faced with a scattered field of competitors, with no incumbent President or heir-apparent to block his way, could use primaries to pick up some delegates and change a few minds.

But no candidate would risk embarrassment by a poor showing in the primaries if an easier pathway to the nomination was open. The rule of thumb among party tacticians was that the earlier a candidate put himself in the field, the weaker the candidacy.[24] As candidate Hubert Humphrey explained to Theodore White during the 1960 primary season, on his campaign bus one bone-chilling Wisconsin winter night:

> You have to be crazy to go into a primary. A primary, now, is worse than the torture of the rack. It's all right to enter a primary by accident, or because you don't know any better, but by forethought . . . .[25]

In 1960 John F. Kennedy's candidacy was regarded as deeply flawed by a number of Democratic party leaders owing to their concerns that a Roman Catholic could not be elected President. Many of these leaders, themselves Roman Catholics, still remembered the doomed presidential candidacy of Alfred E. Smith thirty-two years earlier.[26] And so Kennedy had to take the risks of running in primary elections. As White describes his calculations:

> If he could carry the Tenth [congressional district] he would carry all of Wisconsin. If he swept the entire Wisconsin delegation, he couldn't see how "they" (the big Eastern bosses) could deny him the nomination . . . He mused on the effect a victory here would have on the thinking of Governor Lawrence, boss of vital Pennsylvania . . . .[27]

Of course it was a primary victory, in largely Protestant West Virginia, that more than any other event of 1960 overcame the doubts that party leaders had about the burden of Kennedy's religion.[28] As the results of the general election later demonstrated, they were far from foolish in their apprehensions.[29] But Kennedy's other assets, and the inability of the rest of his competitors to match them, made the risk worth taking.

Kennedy worked hard to reduce his dependency upon party leaders. On one occasion rivalry between leaders within a state party played into his hands. In Ohio, Governor Mike DiSalle wanted to head an uncommitted delegation. But Ray Miller and the Cuyahoga County faction, from Cleveland, threatened to run Kennedy against DiSalle in the primary. As Aaron Wildavsky describes it:

> The astute governor, when he saw how the land lay, entered the primary as a favorite son pledged to Kennedy so long as he had a chance to win the nomination.[30]

In this small event we can catch the beginnings of change: primary elections not merely as an appeal to party leaders but also, and for the time being only in special circumstances, as an appeal against them as well. Primaries had the potential of shifting the balance of initiatives from party leaders and their organizations to candidates and their organizations, if the grip of party leaders was infirm, or their organizations incapable of restricting, or controlling, or dominating a state-wide party electorate, or the candidate in question was especially glamorous, or popular, or a good campaigner, or had friends or allies within the state.

Otherwise, prospective candidates had to calculate carefully whether or not to enter primary elections at all. As V. O. Key described the situation in 1964, in the last edition of his magisterial text:

> An early victory in a pivotal state may win the delegation and impress the party in other states with the aspirant's vote-pulling power. On the other hand, a defeat may bring the boom to a premature end. Yet a refusal to enter the primary may be interpreted as a manifestation of lack of confidence. Since the primary choice may be governed by the wishes of the state organization, it would be rash to enter a primary unless the candidate has the support of the organization or of an important faction, or feels that he could defeat the state machine.[31]

Before 1968, in short, the pursuit of a presidential nomination principally by entering primaries constituted a high-risk strategy. The increasing presence of television, the decline in the influence of political parties on voters generally in many localities, the success of John Kennedy, and the remarkable showing of Estes Kefauver all suggested that it would prove to be more useful to prospective candidates in the years ahead. None of this, however, prefigured the earthquake that was to shake the political system in 1968.

2. The Transformation of 1968

The year 1968 spawned more violence than most years before or since. There were coups in Iraq, Mali, Peru, Sierra Leone, Panama, and Congo (Brazzaville). Congress was suspended in Brazil. There were general strikes in Italy and Pakistan. Severe, protracted riots took place in Poland, Northern Ireland, Mauritius, Mexico, the United Arab Republic, and West Germany. Riots were accompanied by general strikes and demonstrations and the fall of the government in France. Terrorists exploded a bomb in a Jerusalem market, killing eleven people. Czechoslovakia was invaded by the armies of its allies of the Warsaw Pact—notably the Soviet Union—and the Czech leadership that had sponsored a measure of liberalization was forcibly removed. Civil war continued in Nigeria, and there was widespread starvation in what was then known as the secessionist province of Biafra. Universities around the world were visited by demonstrations, disturbances, violence, and disorder and were closed by police or by armies in Mexico City, Paris, and Cairo. The French and the Chinese tested—that is, exploded—thermonuclear devices. The North Vietnamese army and/or the Vietcong launched a major offensive in South Vietnam during the Tet holiday. During the 25-day-long battle of Hue alone an estimated 11,000 people—soldiers and civilians—were killed.

Conditions were very little better for the United States. As a major participant in the war in Vietnam, the United States was greatly affected by the Tet offensive and by the prolongation of the war. During the first six months of 1968, some 9,557 Americans died due to military action, more than the 9,419 Americans who died in Vietnam in the entire year of 1967. By the end of the year, a grand total of more than 30,000 Americans had died since the war began.[32]

In Memphis, Tennessee, Martin Luther King was shot and killed. There followed several days of rioting in more than 120 American cities, with much looting and burning of property. More than forty people lost their lives. The U.S. Ambassador to Guatemala was gunned down, as was presidential candidate Robert F. Kennedy. North Korea seized the U.S. vessel *Pueblo.* Students at Columbia University ransacked and occupied the president of the university's office and closed the school for more than a week. In Miami, Florida, riots broke out during the Republican National Convention in nearby Miami Beach. Three people were killed.

So it went, all year. At least the disorders accompanying the Republican convention took place away from the coverage of television. The Democratic party was not so lucky with its August Chicago convention. There, in full view of television cameras, many ugly incidents of violence took place. These frequently were instigated by peace officers, guards, or the police.

The U.S. government was moved to inaugurate a statistical series based on data collected in the Justice Department. Titled "Civil Disturbances and Related Deaths," it showed 1968 to be a peak year for "major" civil disturbances in the United States.[33] A "major" disturbance was defined as having all of the following characteristics: vandalism, arson, looting or gunfire, the presence of outside police forces or troops, more than 300 participants, excluding police, and a duration of 12 or more hours. Twenty-six such events were recorded in 1968, up from twelve the previous year. Between June of 1967 and June the next year, 148 persons were recorded as having died in civil disturbances. The body count receded to thirty in the following year, and thirty-two in 1969–1970, and then tapered off, as Table 1.2 shows.

Table 1.2   Civil disturbances and related deaths: 1967 to 1973

Period	Total	Disturbances Major¹	Other²	Related deaths	Period	Total	Disturbances Major¹	Other²	Related deaths
1967, June–Oct.	52	12	40	87	1970				
					July–Sept.	20	3	17	6
1968	80	26	54	83	Oct.–Dec.	6	–	6	6
Jan.–Mar.	6	2	4	9					
Apr.–June	46	19	27	52	1971	39	10	29	10
July–Sept.	25	5	20	21	Jan.–Mar.	12	4	8	6
Oct.–Dec.	3	–	3	1	Apr.–June	21	5	16	4
					July–Sept.	5	–	5	–
1969	57	8	49	19	Oct.–Dec.	1	1	–	–
Jan.–Mar.	5	–	5	–					
Apr.–June	27	5	22	8	1972	21	2	19	9
July–Sept.	19	3	16	9	Jan.–Mar.	3	–	3	5
Oct.–Dec.	6	–	6	2	Apr.–June	8	1	7	–
					July–Sept.	5	1	4	1
1970	76	18	58	33	Oct.–Dec.	5	–	5	3
Jan.–Mar.	26	8	18	10					
Apr.–June	24	7	17	11	1973, Jan.–Mar.	4	1	3	2

– Represents zero.   ¹Characterized by all of the following: (a) vandalism; (b) arson; (c) looting or gunfire; (d) outside police forces or troops used; (e) more than 300 persons involved, excluding police; (f) 12 hours or longer duration.   ²Characterized by: Any three elements (a)–(d) described in footnote 1; duration of at least three hours; and more than 50 persons involved, exclusive of police.

Source: U.S. Dept. of Justice, Internal Security Division, unpublished data. *Statistical Abstract of the U.S. 1973* (Washington, D.C.: Government Printing Office, 1973), p. 148.

Not all these events were equally important in supplying a context for the movement toward political reform that culminated in the transformations which provide the occasion of this book. It is important nevertheless to remind ourselves of the atmosphere in which the changes took place. More proximately, three 1968 events were especially significant in shaping the changes as they occurred. They were President Johnson's precipitate withdrawal from the nomination process, the assassination of Robert Kennedy, and the election of Richard Nixon.

3. *The Hasty Departure of President Johnson*

Preliminary plans for the Democratic National Convention were drawn mostly for the convenience of Lyndon Johnson. The convention was to be held in Chicago, Mayor Daley's city, and was timed a shade later than was customary so as to coincide with the President's August 27 birthday.[34] By early in 1968, however, it had become apparent that it would be no ordinary birthday party. A goodly portion of the political activists in the Democratic party were in revolt, open or tacit, against President Johnson's continued prosecution of the Vietnam war. Some were dismayed that it should have become an American war. Others, that there was a Democratic President in charge. Many were convinced of the immorality of the entire enterprise. Some felt they had been deceived at the time of the Tonkin Gulf resolution, which was widely interpreted in Washington as having granted President Johnson broad powers. There was distress at the President's concealment of the actual magnitude of the American commitment and his unwillingness to face the economic costs of the war.[35]

Sentiment out in the country was not much happier. Public opinion surveys told a story of escalating disapproval of the war and of President Johnson's conduct of the Presidency. In early February of 1968 the Gallup poll for the first time showed more respondents agreeing than disagreeing with the proposition that the U.S. had made a "mistake" in getting involved in Vietnam. By March the Gallup poll reported 69 percent in favor of phasing out of Vietnam, with Democrats 70 percent to 18 percent in favor and Republicans 68 percent to 25 percent in favor. President Johnson's overall job rating sank from 48 percent approval and 39 percent disapproval in January 1968 to 41 percent approval and 47 percent disapproval in February.[36]

For some activists in the Democratic party—notably Allard K. Lowenstein of New York—the coming 1968 election seemed an appropriate vehicle for the expression of their deep disapproval of the war and of President Johnson's policy of heavy and increasing American involvement.[37] For the purposes of spreading that message, and of eliciting a response from the electorate, the primary elections seemed ideal. The problem was to find a candidate who would carry the dump-Johnson banner. After some false starts and disappointments,[38] the mantle finally fell on the shoulders of Senator Eugene McCarthy of Minnesota. McCarthy campaigned in New Hampshire, the location of the earliest primary, against the war, against Lyndon Johnson, and against the Presidency itself, arguing that the office had become "personalized" and its powers subject to abuse.[39]

A Johnson write-in movement was hastily organized by regular Democratic politicians in New Hampshire, and it succeeded in beating McCarthy in the March 12 preference primary by 48.5 percent to 42 percent. Far more salient was the virtually unanimous opinion in the news media that McCarthy had pulled off the coup of the year by doing so well.[40]

McCarthy's moral victory emboldened many Democrats around the country to support him and also brought Senator Robert Kennedy of New York into the race. Kennedy in the beginning had refused to oppose the President, and it was only after he had earlier decided against contesting the New Hampshire primary and given his tacit blessing to the effort to find someone else to do the job that McCarthy had agreed to campaign.[41]

The timing of Kennedy's entry into the contest, immediately on the heels of McCarthy's New Hampshire success, did nothing to endear him to McCarthy or to McCarthy's strongest enthusiasts. As one of them, Mary McGrory of the Washington *Star*, wrote:

> Kennedy thinks that American youth belongs to him at the bequest of his brother. Seeing the romance flower between them and McCarthy, he moved with all the ruthlessness of a Victorian father whose daughter has fallen in love with a dustman.[42]

There was a further rationale to Kennedy's candidacy. He was, no doubt in part through the bequest of his brother, more broadly based and more widely acquainted within the Democratic party than McCarthy, notably among black and Hispanic leaders and urban politicians. He was also more widely but not necessarily more favorably known among labor leaders and southern Democratic politicians. There was in addition the sizable cadre of Kennedy loyalists throughout the country who had never been comfortable with Lyndon Johnson or his Presidency and who looked with favor on a restoration of a political status quo that had been interrupted by the assassination of John Kennedy five years before.

As a practical matter, Kennedy people maintained McCarthy could not actually win the nomination whereas Kennedy could. McCarthy was arguing for a drastically scaled-down Presidency, which meant that even if by the unlikeliest turn of events he was nominated and elected he could not or, perhaps worse, would not govern. Kennedy, on the other hand, understood and appreciated the uses of power and could succeed where McCarthy could only fail. Kennedy represented not an anti-establishment but an alternative establishment.[43]

On March 16, Kennedy threw his hat in the ring and entered the next major primary election open to him, the May 7 Indiana primary. Allard Lowenstein's aspiration to make of the primaries a parade of referendums on Lyndon Johnson and his conduct of the Vietnam war was about to be realized.

On March 31, Lyndon Johnson greatly confused the issue by withdrawing as a candidate.[44] This gave the sizable fraction of Democratic party activists that wanted to dump Johnson a stunning victory: Johnson had dumped himself. Still another sizable fraction of the Democratic party remained, however, consisting not only of Johnson loyalists but also of those Democrats who for one reason or another found themselves outside both the McCarthy and the Kennedy camps.

Although taken somewhat by surprise, and inconveniently in Mexico City when the President sprang his announcement, Vice President Hubert Humphrey moved promptly to spread the word of his impending candidacy and announced several weeks later, on April 27.[45] Humphrey's strategy did not include contesting the primary elections. His natural advantage lay in his long-time friendships with party leaders all over the country and especially with labor leaders active in Democratic politics. He wrote:

> Enjoying people as I do, I naturally seek new friends and keep old friendships going. As a matter of course, some of these friendships do have political value, and every politician is human enough to develop some friendships primarily for their potential political value. So I suppose it is honest to say that, within the limitations of my position and the strictures placed on me, I did instinctively try to develop a political base while I was Vice President, building primarily on the base I already had from my senatorial career—my long, happy relationships with the labor movement, the Jewish and black communities, liberal farm organizations. I had also been expanding my contacts in the business world.[46]

While Kennedy and McCarthy were busy campaigning hard against one another in the primary elections, Humphrey pursued a more traditional path:

> An enormous amount of work was done by the key staff people and our labor supporters, the kind of work that is crucial to a solid campaign and is not often seen, because it is not in the forefront: lining up delegates; working the states where delegates were selected in caucuses; getting commitments; putting together the coalition.[47]

This made sense if for no other reason than because so many delegates were not selected in primaries and because as a late entry Humphrey would have had difficulty in securing money and forming a campaign organization quickly. Primaries, moreover, were a risk. Humphrey as Vice President could be burdened with all the disadvantages of having to defend the most unpopular aspects of President Johnson's administration. But as Vice President he had no power to mobilize the Johnson administration or orchestrate events in his own behalf. Journals of opinion—not excluding those liberal organs which had commended him to President Johnson as his vice presidential candidate four years earlier—were already beginning to experiment with the theme that a vote against Humphrey would be a way of striking at the absent Johnson, and that a vote for either Kennedy or McCarthy in a primary election was a repudiation, therefore, of both the President and the Vice President.[48]

There was some evidence that public opinion did not see things quite that way. The Gallup poll of April 28 showed Humphrey, not Kennedy, as the second choice of McCarthy supporters (42% to 31%, with 27% undecided). Only 35 percent of Kennedy supporters, on the other hand, preferred Humphrey. McCarthy was the second choice of 50 percent with only 14 percent undecided.[49]

The climax of the primary season came in early June with what was then the winner-take-all contest for the giant California delegation and its 174 delegate votes. As of May 15, Democratic rank-and-file respondents around the country were split three ways: 40 percent favored Vice President Humphrey, 31 percent were for Senator Kennedy, and 19 percent for Senator McCarthy, with 10 percent undecided.[50] This constituted a nine-point gain for Humphrey over comparable April figures, and losses for Kennedy and McCarthy of four points each. As of the first of June, Democratic county chairmen were overwhelmingly for Humphrey—70 percent Humphrey to 16 percent for Kennedy and 6 percent for McCarthy.[51] Two days before the California primary *Newsweek* claimed 1,279 delegates out of the 1,312 needed to win were solid for or leaning to Humphrey, with 713½ for Kennedy, 280 for McCarthy, and

349 uncommitted.[52] *U.S. News* on the same date put out a much more cautious estimate, with 998 delegates uncommitted or pledged to favorite sons, and Humphrey leading Kennedy by 1,020 delegates to 357, with 247 for McCarthy.[53] Humphrey had campaigned in no primaries and won none. Kennedy had by then won three primaries, and McCarthy had won four.[54] It could reasonably be said on the eve of the California primary election that the Democratic party was split three ways, but with Humphrey comfortably in the lead. His standing with party leaders, with labor, with Johnson loyalists, and in the South, where Kennedy and McCarthy were both notably weak, was considerably greater than either of his main rivals. He was, however, short of victory according to non-partisan calculations, and his rivals were able to make much of the fact that he had not contested any of the primaries, where they had concentrated so much of their effort.

4. The Assassination of Robert Kennedy

What happened next, at approximately 12:20 a.m. on June 5, 1968, was one of the most extraordinary and consequential events in modern American political history. On the night of his victory in the California primary, Robert Kennedy was shot and killed. It was widely and immediately assumed that Vice President Humphrey was the obvious beneficiary of this dreadful act. Events were presently to reveal, however, that nothing could have been farther from the truth.

As Kennedy's friend Dick Tuck disclosed more than a decade later, speaking of the six regional coordinators of the Kennedy campaign who kept track of every delegate and every potential delegate to the 1968 Democratic convention:

> Once their best-kept secret was that Bobby didn't have enough delegates to win the nomination.[55]

Of course it was no secret; most impartial sources put Humphrey far ahead of Kennedy at the time of the California primary.

There was, however, a Kennedy battle plan for the post-primary period: television broadcasts, a twenty-six-state tour, an overseas junket. Humphrey delegate counts were considered soft by Kennedy people, and they thought McCarthy's hold on young people would slip. Of these sorts of wistful thoughts a large and poignant literature has been made.[56] The plausibility of the claim that Kennedy would have been nominated is of less interest than the fact that wishes, made irretrievably contrary to fact by the assassination, for a time dominated the thoughts of those who had been most active in Kennedy's behalf.

As Jules Witcover said:

> For most of the old Robert Kennedy team . . . there was no stomach at this juncture for any more politics in 1968. The pro, Larry O'Brien, hopeful he could help stitch the splintered party together, went over to his friend Humphrey, and [speechwriter Richard] Goodwin returned to McCarthy. But the other key figures . . . sat out the preconvention doldrums . . .

> Part of letting go, among most of the Robert Kennedy followers, was the hope and the anticipation that they could ignore the approaching Democratic National Convention. It would be too painful for them, and it seemed almost irrelevant now . . .

> In all the pulling and hauling [of the convention] over the platform plank on Vietnam, the assassinated candidate's influence was ever-present; in the deep pessimism that infected the convention from the start and inevitably produced a yearning for a viable alternative to Humphrey and McCarthy, the prospect that he could have been nominated, had he lived, seemed infinitely more possible than it had appeared to be on that last night in California.[57]

Although more than a decade has elapsed since the spring of 1968, it is impossible to read the yellowed newspaper clippings, or the outpouring of memoirs from participants in the politics of that difficult time without sensing the intensity of feeling that overcame nearly everyone. Large numbers of Democrats, disgusted with the war and unable to abide the leadership of Lyndon Johnson, had turned to Eugene McCarthy. He in turn had been undercut, as many of his supporters thought, by Rob-

ert Kennedy's unwillingness to endorse their effort and by
Kennedy's belated decision to capture the nomination for himself. Many of Kennedy's people, who had been deeply devoted
to him, were ready enough to adopt a politics of pragmatism
and coalition-building while they were winning, but, overwhelmed by his assassination, afterward turned bitter and de-
spondent.

Humphrey was appalled and immobilized by the assassination of his chief rival:

> Bob Kennedy's death was a national tragedy . . . that shook our
> entire political world. Once more, the poison and hate demon-
> strated in the new act of assassination traumatized the survivors
> and permeated the nation.
>
> To campaign in such a situation was unthinkable, even if one were
> not affected by the act, as I deeply was. This was particularly so
> for a man connected, as I was, with Lyndon Johnson. So I im-
> mediately put a stop to our campaign efforts. We withdrew from
> the New York campaign entirely, even though we'd already spent
> considerably more than any sensible organization should have. We
> lost a month of campaign activity, since it wasn't until July that we
> really geared up once again. Momentum was lost, and now we
> reeled under the strain.[58]

By the end of the pre-convention period, an ugly mood had
settled on the Democratic party. Robert Kennedy's death had
left a gaping wound. Rather than acting as a force for reconcil-
iation Kennedy's partisans joined McCarthy supporters in the
belief that somehow Vice President Humphrey would win the
nomination illegitimately.[59]

The intellectual foundations of this position were set in place
by McCarthy supporters fairly early in the election year. Finding themselves prevented by state party rules from exploiting a
number of opportunities to convert their indignation about the
war into solid delegate votes, McCarthy strategists cried foul.
They put together a blue ribbon commission on the delegate
selection process under the chairmanship of Governor Harold
Hughes of Iowa which served as a clearing house for atrocity

stories about the management of the nomination process by party leaders in various states throughout the country,[60] even in the face of strong challenges by McCarthy and anti-war groups. Among the most telling stories were these:

(1) Unfair representation of McCarthy among the 65 of New York's 190 votes who were appointed by the Democratic State Committee. McCarthy won a slight majority of the 125 elected delegates but received only 15 of the 65 appointed delegates. New York State Chairman John Burns said: "If I ignored the people who have helped me with the organization work and appointed strangers just because they're for McCarthy, I'd have a revolution on my hands."

(2) Walk-outs by McCarthy delegates from state conventions in Connecticut and Kentucky when conventions, controlled by pro-Humphrey majorities, refused to give them what they regarded as adequate representation, in spite of the Humphrey national staff's expressed desire that proportional representation be permitted.

(3) Manipulation by party leaders of access to delegate nomination processes. "In at least two counties in Oklahoma and one congressional district in Missouri, conventions were held in secret. In three Indiana district conventions, McCarthy delegates tape recorded nominating sessions at which the chairman heard the nomination of the regular party slate, entertained a motion to close the nominations and declared an affirmative vote with McCarthy delegates shouting for recognition. One tape records a complete nominating session lasting twenty-two seconds."[61]

Stories of this kind were given wide circulation and fueled resentment of Humphrey and his candidacy by McCarthy supporters, and later by Kennedy supporters. Protests from the Humphrey camp that they had played by the rules were met by many McCarthy and Kennedy partisans with derision. The "old" rules were "old" politics. "New politics" demanded new and more democratic rules.

The Hughes Commission report, *The Democratic Choice*, released on the eve of the Democratic National Convention, concluded that "state systems for selecting delegates to the national convention and the procedures of the convention itself, display considerably less fidelity to basic democratic principles than a nation which claims to govern itself can safely tolerate," and it advocated that an official party commission be appointed to examine problems they had identified. McCarthy delegate Anne Wexler of Connecticut, through her position on the Rules Committee of the Democratic National Convention, was able to bring the commission report to the Committee, which accepted the idea of an ongoing rules commission, and referred Hughes Commission recommendations onward to it.

These complaints raised an issue with which the delegate selection machinery as it existed in 1968 could not cope: what to do about a relatively late-blooming but very strong wave of feeling among a large number of political activists when many delegate positions had already been filled at an earlier time, and when access to the rest of the delegate slots would be hard fought. Not all American political institutions are designed to respond to the sentiments of a given moment. The rationale for electing one-third of the Senate every two years, for example, is expressly to buffer the institution against the effects of a landslide in any one election. So it can hardly be argued as an abstract proposition that responsiveness to contemporary sentiment is an unfailing hallmark of American political institutions. Moreover, support for timely responsiveness to contemporary feelings is not necessarily either a "left-wing" or a "right-wing" policy position. Being for an "open" delegate selection process and favoring McCarthy meant backing timeliness of delegate selection against institutional continuity and was in the context of 1968 politics a more-or-less left-wing position. But to favor up-to-date sentiments a few years later meant supporting the right of state legislatures to overthrow an earlier endorsement of a constitutional amendment favoring the Equal Rights Amendment.

This suggests that a political analysis of real-world consequences underlying the "principles" of timeliness on one hand or of institutional protection from momentary swings of sentiment on the other is appropriate. Emphasis on timeliness in delegate selection gives advantages to candidate enthusiasts and argues for the use of primary elections to select delegates within the calendar year of the election; state party leaders, on the other hand, gain in influence when delegates are selected earlier than the election year and in less public circumstances. So there is a necessary trade-off between the satisfaction of urgent needs for the representation of current feelings and the maintenance of the stability of the party organization. In practice, in 1968, the particular solution to this problem that prevailed in the Democratic party proved unsatisfactory to many of those who sought a reflection of their views, especially about Vietnam, in the delegate selection process. Feelings of having been unjustly treated on the part of some of those who were thus rebuffed were directed against their party's eventual nominee.

Events of the convention week did little to soothe the sensibilities of those outside the Humphrey camp, and the fact that Humphrey was treated little better than they was lost on them. A strenuous effort to write a peace plank on Vietnam for the party platform foundered on the intransigence of the absent Lyndon Johnson. Convention planning was entirely in the hands of Johnson, who rejected Humphrey's plea to move the site from Chicago and refused to cooperate with Humphrey or his allies on technical arrangements of all kinds. Johnson's evident hostility to the McCarthy camp on similar issues—telephones, floor passes, gallery seats—merely convinced McCarthy supporters that the convention was being run, as they told Allen Otten of the *Wall Street Journal,* "as part of a Johnson-Humphrey plot."[62]

The convention itself was tightly managed by large numbers of functionaries in charge of "security" alleged to be taking orders variously from President Johnson, the Secret Service, or Mayor Daley. Numerous disgraceful incidents took place on the

floor of the convention itself or nearby involving the manhan-
dling or arrest of delegates.[63] One delegate was hand-cuffed
and hauled off to jail because he had succeeded in activating a
machine guarding an entrance to the convention hall by using
an ordinary credit card instead of the plastic card issued him
by the Democratic National Committee. Another delegate was
ejected from the floor bodily, in the full view of the television
cameras, because he was alleged to be wearing faulty identity
cards. This occurred over the objection of his delegation chairman, who vouched personally for his identity. Security guards
attacked CBS news reporter Dan Rather and knocked him down
on the floor of the convention hall. The event was broadcast
nationwide, and audiences could hear the most trusted non-
partisan voice in America, that of CBS anchorman Walter
Cronkite, refer to the employees of the Democratic party who
did it as "thugs."[64]

The convention hall was barricaded off from the crowds of
protesters, demonstrators, ill-wishers, and hangers-on who had
come to Chicago for the convention, but the streets and hotels
where delegates stayed were filled with manifestations of their
discontent. These grew greatly in pitch and volume as Mayor
Daley's police herded delegates about and, seemingly at random, attacked people crossing the path of the police.

After it was all over, an official investigation of irregular conduct that took place in the surrounds of the convention yielded
the following description of a part of what went on:

> During the week of the Democratic National Convention, the Chi-
> cago police were the targets of mounting provocation by both word
> and act. It took the form of obscene epithets, and of rocks, sticks,
> bathroom tiles and even human feces hurled at police by demon-
> strators . . . .
>
> The crowd included Yippies come to "do their thing," youngsters
> working for a political candidate, professional people with dis-
> senting political views, anarchists and determined revolutionaries,
> motorcycle gangs, black activists, young thugs, police and secret
> service undercover agents. There were demonstrators waving the

> Viet Cong flag and the red flag of revolution and there were the simply curious who came to watch and, in many cases, became willing or unwilling participants . . . .

> [I]ncidents of intense and indiscriminate violence occurred in the streets . . .

> Demonstrators attacked too. And they posed difficult problems for police as they persisted in marching through the streets, blocking traffic and intersections. But it was the police who forced them out of the park and into the neighborhood.

> On the part of the police there was enough wild club swinging, enough cries of hatred, enough gratuitous beating to make the conclusion inescapable that individual policemen, and lots of them, committed violent acts far in excess of the requisite force for crowd dispersal or arrest. To read dispassionately the hundreds of statements describing at firsthand the events of Sunday and Monday nights is to become convinced of the presence of what can only be called a police riot.[65]

Needless to say, unofficial observers, drawn to Chicago to witness the Democratic convention, and some of them deeply engaged politically, were not always so restrained in their accounts. A particularly ripe specimen—not greatly atypical of much left-Democratic "coverage" of political events of 1968—appeared in the *New York Review of Books* in late September from the pen of the well-known novelist and Eugene McCarthy supporter, William Styron:

> It was perhaps unfortunate that [Mayor Richard J.] Daley, the hoodlum suzerain of the city, became emblematic of all that the young people in their anguish cried out against, even though he plainly deserved it. No one should ever have been surprised that he set loose his battalions against the kids; it was the triumphant end-product of his style, and what else might one expect from this squalid person whose spirit suffused the great city as oppressively as that of some Central American field marshal. . . . It was too bad that Daley should have hogged a disproportionate share of the infamy which has fallen upon the Democratic party; for if it is getting him off the hook too easily to call him a scapegoat, nonetheless the execration he has received . . . may obscure the fact that Daley is only the nastiest symbol of stupidity and desue-

tude in a political party that may die, or perhaps is already dead, because it harbors too many of his breed and mentality. Humphrey, the departed John Bailey, John Connally, Richard Hughes, Muskie—all are merely eminent examples of a rigidity and blindness, a feebleness of thought, that have possessed the party at every level.

Recalling those young citizens for Humphrey who camped out downstairs in my hotel, that multitude of square, seersuckered fraternity boys and country club jocks with butch haircuts, from the suburbs of Columbus and Atlanta, who passed out Hubert buttons and Humphrey mints, recalling them and their elders, mothers and fathers, some of them delegates and not all of them creeps or fanatics by any means but an amalgam of everything— simply well-heeled, most of them, entrenched, party hacks tied to the mob or with a pipeline to some state boss, a substantial number hating the war but hating it not enough to risk dumping Hubert in favor of a vague professorial freak who couldn't feel concern over Prague and hung out with Robert Lowell—I think now that the petrification of a party which allowed such apathy and lack of adventurousness and moral inanition to set in had long ago shaped its frozen logic, determined its fatal choice months before McCarthy or, for that matter, Bobby Kennedy had come along to rock, ever so slightly, the colossal dreamboat. And this can only reinforce what appears to me utterly plausible: that whatever the vigor and force of the dissent, whatever one might say about the surprising strength of support that the minority report received on the floor, a bare but crucial majority of Americans still is unwilling to repudiate the filthy way. This is really the worst thought of all.[66]

A significant feature of many accounts of violence at the Democratic convention was the effort to pin the blame on the Democratic nominee, Hubert Humphrey. On the night of the nomination several of the worst incidents took place, and they were shown on national television, sometimes in embarrassing juxtaposition to coverage of Humphrey's nomination, as Theodore White graphically described:

[San Francisco Mayor Joseph] Alioto rose on screen to nominate him; back and forth the cameras swung from Alioto to pudgy, cigar-smoking politicians, to Daley, with his undershot, angry jaw, painting visually without words the nomination of the Warrior of Joy as a puppet of the old machines. Carl Stokes, the black mayor of Cleveland, was next—to second Humphrey's nomination—and then, at 9:55, NBC's film of the bloodshed had finally been edited, and Stokes was wiped from the nation's vision to show the violence in living color.

The Humphrey staff is furious—Stokes is their signature on the Humphrey civil-rights commitment; and Stokes' dark face is being wiped from the nation's view to show blood—Hubert Humphrey being nominated in a sea of blood.[67]

Amidst all the anger and turmoil of the convention week Humphrey made several efforts to build some bridges to disaffected delegates. In the main, he was ineffectual and, worse, was seen to be buffeted by events rather than in charge of the Democratic party. His efforts to be conciliatory on the Vietnam peace plank of the platform came to naught.[68]

But on one matter, largely unnoticed at the time, Humphrey was able to make a concession unimpeded by President Johnson. This was on the liberal minority report of the Committee on Rules and Order of Business, which used language similar to a resolution adopted with the report of the Credentials Committee and called for abolition of unit voting in state delegations at future national party conventions, and throughout the delegate selection process, and for efforts to assure that in the future "delegates are selected through party primary, convention or committee procedures open to public participation within the calendar year of the National Convention." This report was more or less the result of Hughes Commission recommendations, and unlike the Credentials Committee resolution, called for prompt implementation in the 1972 convention. The Credentials Committee resolution provided only that changes of this sort should be studied by a special commission and a report presented to the 1972 convention.[69] This rather muddled set of overlapping resolutions was the seed from which the most significant party reforms grew, providing the rationale for the formation after the election of a commission of the Democratic party on delegate selection, and for possible implementation of its report in the convention of 1972.

The Commission on Party Structure and Delegate Selection was the formal name given to the McGovern—later the Mc-Govern-Fraser—Commission of the Democratic Party.[70] It was formed in February 1969,[71] and reported to the Democratic National Committee by April 1970. It pieced its own mandate together from the two convention resolutions, one of them calling for "full and timely," the other for "meaningful and timely," opportunities for Democrats to participate in the delegate selection process.[72] Its report, *Mandate for Reform,* contains several elements, notably, (1) endorsements of itself by various Democratic party leaders and strong claims about its legal authority to issue orders to state parties;[73] (2) a compendium of complaints reminiscent of the Hughes Commission about selection procedures for delegates in 1968; and (3) eighteen detailed "guidelines." The importance of these guidelines and their various requirements for subsequent events was very great.[74]

They were put forward, in the words of the report, as ". . . the minimum action state Parties must take to meet the requirements of the Call of the 1972 Convention."[75] Thus, formerly more or less independent state party organizations were put on notice that they would be expected to follow requirements of the national party as outlined by the McGovern Commission. Most of the guidelines dealt with the processes by which delegates were to be chosen. Two selection systems were banned entirely. Byron Shafer describes them as follows:

> *The Party Caucus.* In party caucus systems, the bottom level of party officers—usually precinct committeemen and committeewomen—meet to select delegates to some higher-level convention, which eventually chooses the national convention delegation. States using this system varied according to the level at which the process began and the number of levels which intervened between the bottom and the top. But in toto, the device was the oldest and most widely used delegate selection institution in American history.

The Delegate Primary. In a delegate primary, all candidates for national convention delegate appear on the ballot under their own names only: they may, of course, campaign for election by publicizing their loyalty to a presidential hopeful, but they may also campaign by emphasizing community ties, celebrity, or whatever they prefer. Major states like New York and Pennsylvania had used the device for half a century.[76]

Under the guidelines a state committee would be allowed to select no more than 10 percent of its delegation. This left as approved means of delegate selection the participatory convention and the candidate primary. In Shafer's description:

The Participatory Convention. The structure of a participatory convention system is similar to that of a party caucus, except that participation on the bottom level is open to any party member— and usually to anyone who claims to be a member.

The Candidate Primary. In candidate primaries, the name of the presidential candidate, rather than that of the potential delegate, is featured. In some states, the delegate merely puts his presidential preference after his own name, but in others, only the name of the presidential candidate appears, and the individual delegates whom he wins are actually selected through some other procedure.[77]

The Commission further required that the composition of delegations should include ". . . minority groups, young people and women in reasonable relationship to their presence in the population of the State."[78] The inclusion of ex-officio delegates was entirely prohibited.

How these guidelines were arrived at, how the McGovern-Fraser Commission was staffed and managed, and how those who controlled the Commission succeeded in convincing leaders of the Democratic party to acquiesce in their own broad interpretation of their powers to require changes in state party selection procedures is a thoroughly absorbing and complicated series of stories growing out of the politics of the Democratic party in 1968, but it will not engage us here.[79] Our concern is with the consequences of the undoubted fact that, through in-

ternal processes of the Democratic party and then, finally, in a
court of law,[80] the conclusions of this Commission in due course
became binding upon the Democratic parties of the several
states.

5. *The Contribution of Richard Nixon*

A third signal event of 1968 was the defeat of Vice President
Humphrey in the general election by Richard Nixon. This
meant that there would be no appeal from the work of the
McGovern-Fraser Commission to an incumbent Democratic
President, no court of last resort for those party leaders or state
party organizations or interest groups that felt aggrieved or
disadvantaged by the results of Commission deliberations, no
second opinion on the view adopted by national Democratic
party officials that the Commission had the right to prescribe
uniform rules of behavior for state delegations everywhere.

In addition, Mr. Nixon made a further contribution of enormous later significance by inspiring the Watergate investiga-
tions of the early 70s which led directly to the great changes in
the ways and means by which presidential elections are financed.

The regulation of political finance in the United States has a
long but uninteresting history.[81] It becomes interesting only in
the wake of the extraordinary series of political events that
started with the 1972 break-in at the Democratic National
Committee headquarters in the Watergate office building in
Washington, D.C., by agents of the Committee To Re-elect the
President. The enactment of the Federal Election Campaign
Act Amendments of 1974, the first regulations of the financing
of presidential elections ever to possess teeth, grew directly out
of the Watergate mess. As a distinguished panel of students of
public administration said:

> Few citizens of the United States would disagree with the assertion
> that among the most disturbing revelations of Watergate have been
> those related to the financing and conduct of our federal political

campaigns. An overwhelming majority favors action to prevent recurrence of the scandalous features that infect the system by which money is raised and the uses to which it has been put.[82]

It is not obvious that Watergate was a scandal about political finance at all. If it was, it could be argued that Watergate was about the expenditure of campaign funds for purposes that were already thoroughly illegal, and conceivably also about raising money illegally as well. By this reasoning, no new laws with respect to the regulation of political finance were germane to righting the wrongs of Watergate.

New laws were nevertheless what ensued. It is a common experience for political innovations to be enacted in the wake of situations widely perceived as crises.[83] These innovations are not necessarily rigorously related to the causes of crises or to their amelioration, however. Thus, in the wake of the alarm in the United States about growing technological proficiency in the Soviet Union set off by the launching of Sputnik in October of 1957, the condition of American rural libraries was greatly improved. This was, among other things, an entrepreneurial achievement of a particularly shrewd legislative craftsman in the House of Representatives named Carl Elliott, who understood how an urgent need for Congress to be seen to be doing something could be dovetailed to the requirements of his longer-term political agenda.[84]

No doubt some similar process was at work in producing the 1974 amendments to the Federal Election Campaign Act, which introduced sweeping innovations having broad effects on presidential selection.[85] The new law set limits on the amount of money individuals, political committees, and political parties could contribute to a single election campaign for federal offices.[86] Limits were also set to the amount of money a candidate could spend to get elected. Candidates for a presidential nomination were prohibited from spending more than $10,000,000, and candidates running in the general election were limited to $20,000,000, all as conditions for receiving public finance in election campaigns. The 1974 amendments also developed the rules under which public financing of presidential campaigns, which had been created by the 1971 Act, could be implemented. National party conventions were to be subsidized by public funds. A six-member Federal Election Commission was created. Changes were also made to strengthen campaign finance disclosure laws, in large part by requiring candidates to establish a single central campaign committee, and to report contributions and expenditures in some detail.[87]

The law was challenged in the courts almost immediately after it took effect on January 1, 1975, by a group of political activists from varying parts of the political spectrum.[88] The Supreme Court found some elements of the new scheme unconstitutional: thus, limits on total expenditures by a candidate, limits on independent expenditures in behalf of a candidate, and limits on contributions by a candidate to his own campaign were invalidated.[89] So was the method by which Federal Election Commission members were to be appointed. The Court did, however, uphold limits on individual, political committee, and party contributions. Likewise, the spending limits imposed on recipients of public funds for presidential campaigns were allowed.

Since the Court decision, rendered in early 1976 at the start of a presidential campaign, in effect shut down the new Federal Election Commission, Congress was put in the position of having to rewrite the procedures for appointing commissioners speedily, before the 1976 campaign progressed much further. While they were at it they added a few new amendments to the law.[90] These changes limited individual contributions to political parties[91] and to political committees,[92] and allowed political committees to contribute no more than $15,000 to a political party in a given year.

In sum, the new arrangements provided for in the Federal Campaign Practices Act Amendments of 1974, as modified by the decision of the Supreme Court in *Buckley et al. v. Valeo,* and subsequent laws drastically changed the face of campaign finance. The innovations included (1) provision for federal subsidy of presidential elections, nominating conventions, and primary election campaigns under certain conditions; (2) strict requirements on the reporting by candidates of contributions to and expenditures of their campaigns; (3) limitations on the amounts of money candidates could spend in behalf of their candidacies as a condition of the receipt of federal financing both in presidential primary and general elections and; (4) finally, restrictions on the amounts of money individuals might give to or spend in coordination with (but not independently of) candidates of their choice. This entire panoply of new provisions and regulations was to be presided over by a newly created regulatory agency, the Federal Election Commission.[93] As we shall see, this sizable package of reforms in presidential campaign finance have interacted with reforms directly focused on the presidential selection process to produce wide-ranging consequences.

Appendix

THE OFFICIAL GUIDELINES OF THE COMMISSION ON PARTY STRUCTURE AND DELEGATE SELECTION TO THE DEMOCRATIC NATIONAL COMMITTEE

On November 19 and 20, 1969, the Commission, meeting in open session in Washington, D.C., adopted the following Guidelines for delegate selection.

PART I—INTRODUCTION

The following Guidelines for delegate selection represent the Commission's interpretation of the "full, meaningful, and timely" language of its mandate. These Guidelines have been divided into three general categories.

A. Rules or practices which inhibit access to the delegate selection process—items which compromise full and meaningful participation by inhibiting or preventing a Democrat from exercising his influence in the delegate selection process.

B. Rules or practices which dilute the influence of a Democrat in the delegate selection process, after he has exercised all available resources to effect such influence.

C. Rules and practices which have some attributes of both A and B.

A. Rules or practices inhibiting access:

1. Discrimination on the basis of race, color, creed, or national origin.
2. Discrimination on the basis of age or sex.
3. Voter registration.
4. Costs and fees.
5. Existence of Party rules.

B. Rules or practices diluting influence:

1. Proxy voting.
2. Clarity of purpose.
3. Quorum provisions.
4. Selection of alternates; filling of delegate and alternate vacancies.
5. Unit rule.
6. Adequate representation of political minority views.
7. Apportionment.

40

C. Rules and practices combining attributes of A and B:

1. Adequate public notice.
2. Automatic (ex-officio) delegates.
3. Open and closed processes.
4. Premature delegate selection (timeliness).
5. Committee selection processes.
6. Slate-making.

PART II—THE GUIDELINES

A-1 Discrimination on the basis of race, color, creed, or national origin

The 1964 Democratic National Convention adopted a resolution which conditioned the seating of delegations at future conventions on the assurance that discrimination in any State Party affairs on the grounds of race, color, creed or national origin did not occur. The 1968 Convention adopted the 1964 Convention resolution for inclusion in the Call to the 1972 Convention. In 1966, the Special Equal Rights Committee, which had been created in 1964, adopted six anti-discrimination standards—designated as the "six basic elements"[1]—for the State

---

[1] *Six basic elements, adopted by the Democratic National Committee as official policy statement, January 1968:*

1. All public meetings at all levels of the Democratic Party in each State should be open to all members of the Democratic Party regardless of race, color, creed, or national origin.

2. No test for membership in, nor any oaths of loyalty to, the Democratic Party in any State should be required or used which has the effect of requiring prospective or current members of the Democratic Party to acquiesce in, condone or support discrimination on the grounds of race, color, creed, or national origin.

3. The time and place for all public meetings of the Democratic Party on all levels should be publicized fully and in such a manner as to assure timely notice to all interested persons. Such meetings must be held in places accessible to all Party members and large enough to accommodate all interested persons.

4. The Democratic Party, on all levels, should support the broadest possible registration without discrimination on grounds of race, color, creed or national origin.

5. The Democratic Party in each State should publicize fully and in such manner as to assure notice to all interested parties a full description of the legal and practical procedures for selection of Democratic Party Officers and rep-

Parties to meet. These standards were adopted by the Democratic National Committee in January 1968 as its official policy statement.

These actions demonstrate the intention of the Democratic Party to ensure a full opportunity for all minority group members to participate in the delegate selection process. To supplement the requirements of the 1964 and 1968 Conventions, the Commission requires that:

1. State Parties add the six basic elements of the Special Equal Rights Committee to their Party rules and take appropriate steps to secure their implementation;

2. State Parties overcome the effects of past discrimination by affirmative steps to encourage minority group participation, including representation of minority groups on the national convention delegation in reasonable relationship to the group's presence in the population of the State.[2]

A-2   Discrimination on the basis of age or sex

The Commission believes that discrimination on the grounds of age or sex is inconsistent with full and meaningful opportunity to participate in the delegate selection process. Therefore, the Commission requires State Parties to eliminate all vestiges of discrimination on these grounds. Furthermore, the Commission requires State Parties to overcome the effects of past discrimination by affirmative steps to encourage representation on the national convention delegation of young people—defined as people of not more than thirty nor less than eighteen years of age—and women in reasonable relationship to their presence in the population of the State.[2] Moreover, the Commission

   resentatives on all levels. Publication of these procedures should be done in such fashion that all prospective and current members of each State Democratic Party will be fully and adequately informed of the pertinent procedures in time to participate in each selection procedure at all levels of the Democratic Party organization.

6. The Democratic Party in each State should publicize fully and in such manner as to assure notice to all interested parties a complete description of the legal and practical qualifications for all officers and representatives of the State Democratic Party. Such publication should be done in timely fashion so that all prospective candidates or applicants for any elected or appointed position within each State Democratic Party will have full and adequate opportunity to compete for office."

[2] It is the understanding of the Commission that this is not to be accomplished by the mandatory imposition of quotas.

requires State Parties to amend their Party rules to allow and encourage any Democrat of eighteen years or more to participate in all party affairs.

When State law controls, the Commission requires State Parties to make all feasible efforts to repeal, amend, or otherwise modify such laws to accomplish the stated purpose.

A-3 Voter registration

The purpose of registration is to add to the legitimacy of the electoral process, not to discourage participation. Democrats do not enjoy an opportunity to participate fully in the delegate selection process in States where restrictive voter registration laws and practices are in force, preventing their effective participation in primaries, caucuses, conventions and other Party affairs. These restrictive laws and practices include annual registration requirements, lengthy residence requirements, literacy tests, short and untimely registration periods, and infrequent enrollment sessions.

The Commission urges each State Party to assess the burdens imposed on a prospective participant in the Party's delegate selection processes by State registration laws, customs and practices, as outlined in the report of the Grass Roots Subcommittee of the Commission on Party Structure and Delegate Selection, and use its good offices to remove or alleviate such barriers to participation.

A-4 Costs and fees; petition requirements

The Commission believes that costs, fees, or assessments and excessive petition requirements made by State law and Party rule or resolutions impose a financial burden on (1) national convention delegates and alternates; (2) candidates for convention delegates and alternates; and (3) in some cases, participants. Such costs, fees, assessments or excessive petition requirements discouraged full and meaningful opportunity to participate in the delegate selection process.

The Commission urges the State Parties to remove all costs and fees involved in the delegate selection process. The Commission requires State Parties to remove all excessive costs and fees, and to waive all nominal costs and fees when they would impose a financial strain on any Democrat. A cost or fee of more than $10 for all stages of the delegate selection process is deemed excessive. The Commission requires State Parties to remove all mandatory assessments of delegates and alternates.

The Commission requires State Parties to remove excessive petition requirements for convention delegate candidates of presidential candidates. Any petition requirement, which calls for a number of signatures in excess of 1% of the standard used for measuring Democratic strength, whether such standard be based on the number of Democratic votes cast for a specific office in a previous election or Party enrollment figures, is deemed excessive.

When State law controls any of these matters, the Commission requires State Parties to make all feasible efforts to repeal, amend or otherwise modify such laws to accomplish the stated purpose.

This provision, however, does not change the burden of expenses borne by individuals who campaign for and/or serve as delegates and alternates. Therefore, the Commission urges State Parties to explore ways of easing the financial burden on delegates and alternates and candidates for delegate and alternate.

A-5   Existence of Party rules

In order for rank-and-file Democrats to have a full and meaningful opportunity to participate in the delegate selection process, they must have access to the substantive and procedural rules which govern the process. In some States the process is not regulated by law or rule, but by resolution of the State Committee and by tradition. In other States, the rules exist, but generally are inaccessible. In still others, rules and laws regulate only the formal aspects of the selection process (e.g., date and place of the State convention) and leave to Party resolution or tradition the more substantive matters (e.g., intrastate apportionment of votes; rotation of alternates; nomination of delegates).

The Commission believes that any of these arrangements is inconsistent with the spirit of the Call in that they permit excessive discretion on the part of Party officials, which may be used to deny or limit full and meaningful opportunity to participate. Therefore, the Commission requires State Parties to adopt and make available readily accessible statewide Party rules and statutes which prescribe the State's delegate selection process with sufficient details and clarity. When relevant to the State's delegate selection process, explicit written Party rules and procedural rules should include clear provisions for: (1) the apportionment of delegates and votes within the State; (2) the allocation of fractional votes, if any; (3) the selection and responsibilities of convention committees; (4) the nomination of delegates and alternates; (5) the succession of alternates to delegate status and the filling of vacancies; (6) credentials challenges; (7) minority reports.

Furthermore, the Commission requires State Parties to adopt rules which will facilitate maximum participation among interested Democrats in the processes by which National Convention delegates are selected. Among other things, these rules should provide for dates, times, and public places which would be most likely to encourage interested Democrats to attend all meetings involved in the delegate selection process.

The Commission requires State Parties to adopt explicit written Party rules which provide for uniform times and dates of all meetings involved in the delegate selection process. These meetings and events include caucuses, conventions, committee meetings, primaries, filing deadlines, and Party enrollment periods. Rules regarding time and date should be uniform in two senses. First, each stage of the delegate selection process should occur at a uniform time and date throughout the State. Second, the time and date should be uniform from year to year. The Commission recognizes that in many parts of rural America it may be an undue burden to maintain complete uniformity, and therefore exempts rural areas from this provision so long as the time and date are publicized in advance of the meeting and are uniform within the geographic area.

B-1 Proxy voting

When a Democrat cannot, or chooses not to, attend a meeting related to the delegate selection process, many States allow that person to authorize another to act in his name. This practice—called proxy voting—has been a significant source of real or felt abuse of fair procedure in the delegate selection process.

The Commission believes that any situation in which one person is given the authority to act in the name of the absent Democrat, on any issue before the meeting, gives such person an unjustified advantage in affecting the outcome of the meeting. Such a situation is inconsistent with the spirit of equal participation. Therefore, the Commission requires State Parties to add to their explicit written rules provisions which forbid the use of proxy voting in all procedures involved in the delegate selection process.

B-2 Clarity of purpose

An opportunity for full participation in the delegate selection process is not meaningful unless each Party member can clearly express his preference for candidates for delegates to the National Convention, or for those who will select such delegates. In many States, a Party member who wishes to affect the selection of the delegation must do so by voting for delegates or Party officials who will engage in many activities unrelated to the delegate selection process.

Whenever other Party business is mixed, without differentiation, with the delegate selection process, the Commission requires State Parties to make it clear to voters how they are participating in a process that will nominate their Party's candidate for President. Furthermore, in States which employ a convention or committee system, the Commission requires State Parties to clearly designate the delegate selection procedures as distinct from other Party business.

B-3   Quorum provisions

Most constituted bodies have rules or practices which set percentage or number minimums before they can commence their business. Similarly, Party committees which participate in the selection process may commence business only after it is determined that this quorum exists. In some States, however, the quorum requirement is satisfied when less than 40% of committee members are in attendance.

The Commission believes a full opportunity to participate is satisfied only when a rank-and-file Democrat's representative attends such committee meetings. Recognizing, however, that the setting of high quorum requirements may impede the selection process, the Commission requires State Parties to adopt rules setting quorums at not less than 40% for all party committees involved in the delegate selection process.

B-4   Selection of alternates; filling of delegate and alternate vacancies

The Call to the 1972 Convention requires that alternates be chosen by one of the three methods sanctioned for the selection of delegates— i.e., by primary, convention or committee. In some States, Party rules authorize the delegate himself or the State Chairman to choose his alternate. The Commission requires State Parties to prohibit these practices—and other practices not specifically authorized by the Call— for selecting alternates.

In the matter of vacancies, some States have Party rules which authorize State Chairmen to fill all delegate and alternate vacancies. This practice again involves the selection of delegates or alternates by a process other than primary, convention or committee. The Commission requires State Parties to prohibit such practices and to fill all vacancies by (1) a timely and representative Party committee; or (2) a reconvening of the body which selected the delegate or alternate whose seat is vacant; or (3) the delegation itself, acting as a committee.

When State law controls, the Commission requires State Parties to make all feasible efforts to repeal, amend or otherwise modify such laws to accomplish the stated purposes.

B-5   Unit rule

In 1968, many States used the unit rule at various stages in the processes by which delegates were selected to the National Convention. The 1968 Convention defined unit rule,[3] did not enforce the unit rule on any delegate in 1968, and added language to the 1972 Call requiring that "the unit rule not be used in any stage of the delegate selection process." In light of the Convention action, the Commission requires State Parties to add to their explicit written rules provisions which forbid the use of the unit rule or the practice of instructing delegates to vote against their stated preferences at any stage of the delegate selection process."[4]

B-6   Adequate representation of minority views on presidential candidates at each stage in the delegate selection process

The Commission believes that a full and meaningful opportunity to participate in the delegate selection process is precluded unless the presidential preference of each Democrat is fairly represented at all levels of the process. Therefore, the Commission urges each State Party to adopt procedures which will provide fair representation of minority views on presidential candidates and recommends that the 1972 Con-

---

[3] UNIT RULE. "This Convention will not enforce upon any delegate with respect to voting on any question or issue before the Convention any duty or obligation which said delegate would consider to violate his individual conscience. As to any legal, moral or ethical obligation arising from a unit vote or rule imposed either by State law by a State convention or State committee or primary election of any nature, or by a vote of a State delegation, the Convention will look to each individual delegate to determine for himself the extent of such obligation if any."

[4] It is the understanding of the Commission that the prohibition on instructed delegates applies to favorite-son candidates as well.

vention adopt a rule requiring State Parties to provide for the representation of minority views to the highest level of the nominating process.

The Commission believes that there are at least two different methods by which a State Party can provide for such representation. First, in at-large elections it can divide delegate votes among presidential candidates in proportion to their demonstrated strength. Second, it can choose delegates from fairly apportioned districts no larger than congressional districts.

The Commission recognizes that there may be other methods to provide for fair representation of minority views. Therefore, the Commission will make every effort to stimulate public discussion of the issue of representation of minority views on presidential candidates between now and the 1972 Democratic National Convention.

B-7   Apportionment

The Commission believes that the manner in which votes and delegates are apportioned within each State has a direct bearing on the nature of participation. If the apportionment formula is not based on Democratic strength and/or population the opportunity for some voters to participate in the delegate selection process will not be equal to the opportunity of others. Such a situation is inconsistent with a full and meaningful opportunity to participate.

Therefore, the Commission requires State Parties which apportion their delegation to the National Convention to apportion on a basis of representation which fairly reflects the population and Democratic strength within the State. The apportionment is to be based on a formula giving equal weight to total population and to the Democratic vote in the previous presidential election.

The Commission requires State Parties with convention systems to select at least 75% of their delegations to the National Convention at congressional district or smaller unit levels.

In convention or committee systems, the Commission requires State Parties to adopt an apportionment formula for each body actually selecting delegates to State, district and county conventions which is based upon population and/or some measure of Democratic strength. Democratic strength may be measured by the Democratic vote in the preceding presidential, senatorial, congressional or gubernatorial election, and/or by party enrollment figures.

When State law controls, the Commission requires State Parties to make all feasible efforts to repeal, amend or otherwise modify such laws to accomplish the stated purpose.

C-1 Adequate public notice

The Call to the 1968 convention required State Parties to assure voters an opportunity to "participate fully" in party affairs. The Special Equal Rights Committee interpreted this opportunity to include adequate public notice. The Committee listed several elements—including publicizing of the time, places and rules for the conduct of all public meetings of the Democratic Party and holding such meetings in easily accessible places—which comprise adequate public notice. These elements were adopted by the Democratic National Committee in January 1968 as its official policy statement and are binding on the State Parties.

Furthermore, the Commission requires State Parties to circulate a concise and public statement in advance of the election itself of the relationship between the party business being voted upon and the delegate selection process.

In addition to supplying the information indicated above, the Commission believes that adequate public notice includes information on the ballot as to the presidential preference of (1) candidates or slates for delegate or (2) in the States which select or nominate a portion of the delegates by committees, candidates or slates for such committees.

Accordingly, the Commission requires State Parties to give every candidate for delegate (and candidate for committee, where appropriate) the opportunity to state his presidential preferences on the ballot at each stage of the delegate selection process. The Commission requires the State Parties to add the word "uncommitted" or like term on the ballot next to the name of every candidate for delegate who does not wish to express a presidential preference.

When State law controls, the Commission requires the State Parties to make all feasible efforts to repeal, amend or otherwise modify such laws to accomplish the stated purposes.

C-2 Automatic (ex-officio) delegates (see also C-4)

In some States, certain public or Party officeholders are delegates to county, State and National Conventions by virtue of their official position. The Commission believes that State laws, Party rules and Party resolutions which so provide are inconsistent with the Call to the 1972 Convention for three reasons:

1. The Call requires all delegates to be chosen by primary, convention or committee procedures. Achieving delegate status by virtue of public or Party office is not one of the methods sanctioned by the 1968 Convention.

2. The Call requires all delegates to be chosen by a process which begins within the calendar year of the Convention. Ex-officio delegates usually were elected (or appointed) to their positions before the calendar year of the Convention.

3. The Call requires all delegates to be chosen by a process in which all Democrats have a full and meaningful opportunity to participate. Delegate selection by a process in which certain places on the delegation are not open to competition among Democrats is inconsistent with a full and meaningful opportunity to participate.

Accordingly, the Commission requires State Parties to repeal Party rules or resolutions which provide for ex-officio delegates. When State law controls, the Commission requires State Parties to make all feasible efforts to repeal, amend or otherwise modify such laws to accomplish the stated purpose.

C-3 Open and closed processes

The Commission believes that Party membership, and hence opportunity to participate in the delegate selection process, must be open to all persons who wish to be Democrats and who are not already members of another political party; conversely, a full opportunity for all Democrats to participate is diluted if members of other political parties are allowed to participate in the selection of delegates to the Democratic National Convention.

The Commission urges State Parties to provide for party enrollment that (1) allows non-Democrats to become Party members, and (2) provides easy access and frequent opportunity for unaffiliated voters to become Democrats.

C-4 Premature delegate selection (timeliness)

The 1968 Convention adopted language adding to the Call to the 1972 Convention the requirement that the delegate selection process must begin within the calendar year of the Convention. In many States, Governors, State Chairmen, State, district and county committees who are chosen before the calendar year of the Convention, select—or choose agents to select—the delegates. These practices are inconsistent with the Call.

The Commission believes that the 1968 Convention intended to prohibit any untimely procedures which have any direct bearing on the processes by which National Convention delegates are selected. The process by which delegates are nominated is such a procedure. Therefore, the Commission requires State Parties to prohibit any practices by which officials elected or appointed before the calendar year choose nominating committees or propose or endorse a slate of delegates— even when the possibility for a challenge to such slate or committee is provided.

When State law controls, the Commission requires State Parties to make all feasible efforts to repeal, amend, or modify such laws to accomplish the stated purposes.

C-5 Committee selection processes

The 1968 Convention indicated no preference between primary, convention, and committee systems for choosing delegates. The Commission believes, however, that committee systems by virtue of their indirect relationship to the delegate selection process, offer fewer guarantees for a full and meaningful opportunity to participate than other systems.

The Commission is aware that it has no authority to eliminate committee systems in their entirety. However, the Commission can and does require State Parties which elect delegates in this manner to make it clear to voters at the time the Party committee is elected or appointed that one of its functions will be the selection of National Convention delegates.

Believing, however, that such selection system is undesirable even when adequate public notice is given, the Commission requires State Parties to limit the National Convention delegation chosen by committee procedures to not more than 10 percent of the total number of delegates and alternates.

Since even this obligation will not ensure an opportunity for full and meaningful participation, the Commission recommends that State Parties repeal rules or resolutions which require or permit Party commit-

52

tees to select any part of the State's delegation to the National Convention. When State law controls, the Commission recommends that State Parties make all feasible efforts to repeal, amend, or otherwise modify such laws to accomplish the stated purpose.

C-6 Slate-making

In mandating a full and meaningful opportunity to participate in the delegate selection process, the 1968 Convention meant to prohibit any practice in the process of selection which made it difficult for Democrats to participate. Since the process by which individuals are nominated for delegate positions and slates of potential delegates are formed is an integral and crucial part of the process by which delegates are actually selected, the Commission requires State Parties to extend to the nominating process all guarantees of full and meaningful opportunity to participate in the delegate selection process. When State law controls, the Commission requires State Parties to make all feasible efforts to repeal, amend or otherwise modify such laws to accomplish the stated purpose.

Furthermore, whenever slates are presented to caucuses, meetings, conventions, committees, or to voters in a primary, the Commission requires State Parties to adopt procedures which assure that:

1. the bodies making up the slates have been elected, assembled, or appointed for the slate-making task with adequate public notice that they would perform such task;

2. those persons making up each slate have adopted procedures that will facilitate widespread participation in the slate-making process, with the proviso that any slate presented in the name of a presidential candidate in a primary State be assembled with due consultation with the presidential candidate or his representative;

3. adequate procedural safeguards are provided to assure that the right to challenge the presented slate is more than perfunctory and places no undue burden on the challengers.

When State law controls, the Commission requires State Parties to make all feasible efforts to repeal, amend or otherwise modify such laws to accomplish the stated purpose.

II

Consequences for Political Parties

1. Changing the Structure of Incentives

A large body of commentary now exists discussing the consequences for parties of party reform. Those critical of the reforms have argued the view that they brought significant unanticipated changes to American politics;[1] those more favorable have argued that they have had a far more limited impact, except as intended.[2] It may not be possible to find empirical grounds for settling all the disagreements about the effects of party reform in a definitive way, but it is, perhaps, possible to make headway in establishing the plausibility of at least some of the details on which the contrary positions rest.

In general the argument proceeds mostly through the discussion of what happened in the Democratic party. In many respects, the Republican party remains unreformed: its apportionment formula for delegates to national conventions is still weighted toward the electoral college; the 1976 recommendations of its Rule 29 Committee to provide for monitoring of state delegation selection processes were rejected by the Republican National Committee and the 1976 Convention; the Republicans have no enforceable demographic quotas or affirmative action standards; convention committees reflect state equality in apportionment of members; and the confederate legal structure of the party has been retained.[3] Yet important consequences of the reforms of the Democratic party have been visited upon the Republicans as well; the laws having to do with campaign finance operate equally on both parties, and so do the effects of the mass media. The impulse to reform the nomination process came out of the turmoil of the Democratic experience of 1968, but because the delegate selection procedures disapproved of by the McGovern-Fraser guidelines in many states were perfectly legal and established in state law, changes in the election laws of the several states were necessary in order for many Democratic state parties to come into compliance with the demands of the emerging enforcement machinery of the Democratic National Committee. Where state laws were changed, willy-nilly, Republican practices had to be adjusted as well, presumably because changes in the rules under which delegate selection was to take place changed the structure of opportunities for Republican as well as for Democratic politicians.

Thus the notion that party reforms have had a significant impact upon the parties relies upon the argument that party leaders in the various states responded to the imposition of McGovern-Fraser guidelines and to other changes in the political climate by modifying their own behavior in accord with a newly emerging set of incentives.

At least five such incentives can be identified, all working in mutually reinforcing ways to transform the presidential nominating process. In brief, they are: (1) centralization of control over the certification of delegates, arising from the uncertainties introduced by the McGovern Commission guidelines; (2) federal subsidies for candidates to contest primaries, interacting with the mass media and with the public commitment of delegates to candidates so as to push critical decisions earlier and earlier on the calendar; (3) restriction of sources of money other than the subsidy; (4) changes in primary election rules promoting ease of candidate entry; and (5) the separation of state conventions from delegate selection procedures owing to the risks of contamination of state conventions by candidate enthusiasts. The first and last of these have affected the leaders of state parties, inducing them to establish primaries for candidates to contest, while the middle three chiefly affected presidential candidates, enticing them into entering primary elections.

(1) The promulgation by the McGovern Commission of its eighteen guidelines introduced a new order of complexity, of uncertainty, and of centralized control to the selection of state delegations to national party conventions. Complexity meant the possibility of widely differing interpretations of what constituted adequate compliance with guidelines, and the possibility of conflict over the issue of compliance within the state parties. Uncertainty meant that state party leaders could not be sure whether the full range of options theoretically open to them in designing a presidential selection process would in fact pass muster with the compliance machinery being evolved by the Democratic National Committee or with the credentials committee of the next national convention. Neither body could be presumed in advance by state party leaders to be purely ministerial in character, devoid of political coloration. Centralization meant that as far as state party leaders were concerned, the ultimate determination of the adequacy of their compliance would be out of their hands. The burden of proof concerning a state delegation's fitness to be seated at a national convention had been shifted onto the shoulders of state party leaders, and this left them vulnerable to possible challengers within the state to the delegation's credentials before the national convention credentials committee.[4]

As the 1972 delegate selection season approached, the uncertainties of party leaders were palpable. R. W. Apple of the *New York Times* wrote:

> The delegates will have been chosen in a bewildering variety of ways—but all or nearly all by methods sanctioned by the reform

commission headed during the months of its most active work by
Senator George McGovern of South Dakota. . . .

In 22 states and the District of Columbia, the delegates will be
selected in some form of primary election.

In 24 states, a convention system with procedures already set will
be used. In four others, a convention system will also be used, in
all probability, but the situation is too murky at the moment to
say for sure.

Delaware is one of the places where confusion reigns (the others
are Georgia, Louisiana and Michigan). Under the rules as they
now stand, delegates to the state convention, which would pick the
delegates to the national convention, would not even be chosen
until July 8, two days before the national convention is scheduled
to begin. . . .

In all, more than 60 per cent of the delegates will be elected in
primaries. . . .

But the McGovern commission requirement that at least 75 per
cent of a state's delegation be elected by Congressional district (to
avoid submerging minorities) forces more complicated systems on
most convention states. The only exceptions are the six states that
have only one Congressional district. . . .

The minutiae of election law and party regulations are usually of
interest only to political insiders. But they can influence major
political decisions.

For example, one of Governor [John] Gilligan's main reasons for
coming out for Senator Muskie, rather than simply throwing the
state open to all comers, was the fact that the Ohio election law
requires all candidates for delegate to state a preference. All Ohio
delegates will be chosen in the primary. So if Mr. Gilligan did not
want to run as a favorite son, he either had to state his choice or
give up any plan to attend the convention.[5]

Selecting delegates by primary election offered a quick and
simple solution to most of the problems state party leaders were
experiencing since it was evidently widely presumed that a primary conducted in the usual way would almost certainly meet
the requirements of the guidelines.[6]
It has been denied that the proliferation of primaries in the

delegate selection process—from 17 Democratic and 16 Republican in 1968, before the guidelines were promulgated, to 23 Democratic and 22 Republican in 1972, the next presidential election year—was inspired by McGovern-Fraser Commission reforms. Two former staff members of the Commission—conceivably in a mood to defend their work from well-meant erosion or from political attack—have generated a short list of other possible motives for switching to primary elections.[7] These include a desire on the part of state party leaders to attract the coverage of the national television networks for the purposes of drawing attention to state or regional problems. In addition, there is the simple financial attraction of having a primary, since:

> The candidates, their campaigns, and the press who cover them are known to spend a great deal of money in "crucial" primary states.[8]

These authors also mention several states—Texas, North Carolina, Georgia—in which they believe that primary elections were instituted because such elections were going to be helpful in 1976 to the favorite son presidential candidacies of Lloyd Bentsen, Terry Sanford, and Jimmy Carter, respectively. This ignores the changes of the 1968–72 period, but raises the possibility that by the time the leaders of state parties began thinking about the 1976 elections a whole raft of reasons to switch to primary elections might have occurred to them. It is not entirely clear, however, why primary elections were necessarily superior to the methods of delegate selection they replaced, at least from the standpoint of favorite son candidates. In Georgia, for example, prior to the reforms "the entire national convention delegation was chosen by the state party chairman, in consultation with the Democratic governor."[9] Even though the Georgia state chairman in 1976 was no friend of Jimmy Carter's, it is easy to envisage such a system being even more advantageous than a primary to a favorite son. If being selected by delegates picked in a primary election rather than by other means could be regarded as particularly helpful to candidates

Table 2.1 Major changes in delegate selection procedures: 1968–72, 1972–76, 1976–80

State	1968	1972
Illinois	direct primary, caucus	clsd pref primary
Maryland	(d) committee, (r) caucus	clsd pref primary
Michigan	(d) caucus	(d) clsd pref primary
New Mexico	caucus	clsd pref primary
North Carolina	caucus	clsd pref primary
Pennsylvania	clsd pref primary, committee	clsd pref primary
Rhode Island	(d) committee, (r) caucus	clsd pref primary (+ *inds*)
Tennessee	caucus	open primary

State	1972	1976
Arkansas	caucus	open primary
Georgia	caucus	open primary
Idaho	caucus	open primary
Kentucky	caucus	clsd pref primary
Michigan	(r) caucus	(r) clsd pref primary
Montana	(d) caucus	(d) clsd pref primary
Nevada	caucus	clsd pref primary
New Mexico	clsd pref primary	caucus
Texas	caucus	open primary

State	1976	1980
Arkansas	(r) open primary	(r) caucus
Connecticut	caucus	clsd pref primary
Idaho	(d) open primary	(d) caucus
Kansas	caucus	clsd pref primary (+ *inds*)
Louisiana	caucus	clsd pref primary
Michigan	(d) open primary	(d) caucus
Mississippi	(r) caucus	(r) direct primary
New Mexico	caucus	clsd pref primary
New York	(d) direct primary, committee	(d) direct primary
South Carolina	(r) caucus	(r) open primary
Puerto Rico	caucus	open primary

(r) Republicans only; (d) Democrats only; (inds) open to Independents as well; *caucus,* Delegates chosen by state and local caucuses and conventions; *committee,* Delegates chosen by state party committee; *clsd pref primary,* Delegates chosen or bound by presidential preference primaries open only to voters preregistered as members of the particular party; *direct primary,* Delegates chosen directly by voters in primaries, without a presidential preference poll; *open primary,* Delegates chosen or bound by presidential preference primaries open to all registered voters without regard to party pre-registration.

Sources: Austin Ranney (ed.), *The American Elections of 1980* (Washington, D.C.: American Enterprise Institute, 1981), pp. 366–68, and *Congressional Quarterly Weekly Reports*.

because delegates picked in this fashion would be perceived to be more legitimate by press and public, or less vulnerable to credentials challenges, then of course one must ask why. The answer seems plain enough: because of the changes in party regulations inspired by the McGovern Commission.

The other reasons given make sense as possible excuses for switching to primary elections at any time; what they do not explain is why all these reasons so suddenly occurred to so many state party leaders, and only after the guidelines had come into being. Could the compulsion of newly instituted guidelines have coincidentally stimulated an interest in publicizing state problems, or in raising money for the tourist industry? Perhaps; but it is more likely that the Winograd Commission of the Democratic Party—a commission set up after the 1976 election to examine the effects of reform—was expressing the view of many state party leaders when it said that "many states . . . felt that a primary offered the most protection against a challenge at the next convention."[10] The primary election was in any event a solution that was adopted in most cases where states changed from one election to the next in the period after 1968.

(2) The creation, by the Federal Election Campaign Act Amendments of 1974, of a federal subsidy for candidates in primary elections encouraged candidates to enter primaries.[11] In order to qualify for the subsidy, candidates had to raise $5,000 or more in small amounts (the federal subsidy only matches each donor's first $250) in twenty different states, a requirement that has not turned out to be onerous.

This subsidy also creates an objective device through which journalists can sift the various candidates, declaring those that have qualified for the subsidy as "serious" contenders. Although the actual payments of the subsidy do not occur until the election year begins, slowness in acquiring what Terry Sanford referred to as this "license to practice" could restrict the amount of favorable news coverage a candidate might receive.[12]

Not only participation in primaries but in early primaries is thus encouraged. The media coverage attending announcements of participation in early primaries has provided a means for qualifying for the subsidy, since the publicity surrounding entry into the race can help to generate the name recognition necessary to inspire a sufficient number of recipients of mailed appeals for funds to contribute a small amount, thus activating the matching federal payment. A further incentive to early entry into primaries is created because early choices color the way later alternatives are perceived. R. W. Apple said in 1972:

> Because the process in Iowa and in most of the other convention states is so structured and so open to public view, it appears likely that, beginning at the lowest level, the preferences of delegates will be known. From step to step, it will be relatively easy to assess Mr. Muskie's strength, Mr. McGovern's and so on.

> Thus, it will be far harder than in the past to remain uncommitted. Some analysts believe that fewer than 15 per cent of the delegates at Miami Beach will actually be uncommitted.[13]

As this suggests, the news media have learned to give enormous coverage to, and tend to extract a coherent story from the results of, the earliest primaries and state conventions.[14] More delegates are bound to candidates by rules prohibiting delegate primaries, and "uncommitted" delegates selected by state conventions tend to be ignored in news stories, as happened most conspicuously in the coverage of Iowa in 1976. The strategic options for political leaders, candidates, and party influentials alike are increasingly foreclosed by the publicity blitz of the early primaries, and the competitive pressure among newspapers, magazines, and the networks for reaction stories. Thus early participation becomes a necessity for a serious candidate. As Michael Robinson says, "In proportionate terms, each Democratic vote [in 1976] in New Hampshire received 170 times as much network news time as each Democratic vote in New York. Media reality—television reality—implied that a victory in New Hampshire totally overwhelmed a victory in New York."[15]

Candidates must allocate as well as raise their money early because of the way early results influence later outcomes. As reports filed with the Federal Election Commission show, this leads to great disparities in the investments candidates make in the different state selection processes, depending on the timing of delegate selection.

Table 2.2 Amount of money candidates receiving public financing spent per vote in selected states

State (Delegates selected in)	1976	1980
Iowa (Jan.)	$9.46	$13.89
New Hampshire (Feb.)	7.22	8.90
Florida (March)	2.02	2.21
Wisconsin (April)	1.15	1.36
Nebraska (May)	1.30	.45
California (June)	.68	.23

Source: *Congressional Quarterly Weekly Report* (December 26, 1981), p. 2566. See also F. Christopher Arterton, "Dollars for Presidents: Spending Money under the FECA," in Campaign Finance Study Group, *Financing Presidential Campaigns* (Mimeo), (Institute of Politics, John F. Kennedy School of Government, Harvard University, January 1982), chap. 3, pp. 11–12.

(3) Candidates were also encouraged to enter primaries, seek early publicity and qualify for the federal subsidy because access to the principal alternative source of funds to run any sort of campaign at all—heavy financial backing by a small circle of friends—was after 1974 forbidden by law.[16] Thus the principal sources of funds outside the federal subsidy itself became small individual contributions: money raised by direct-mail advertising, through anonymous donations at large fund-raising events like rock concerts, at $250-a-plate dinners, and similar activities.[17]

(4) Candidates were even further encouraged to enter early primaries by the abolition of the unit rule at the 1968 Democratic convention and consequently of winner-take-all primary elections. This meant that several candidates in any given election might anticipate garnering at least some pledged delegates, that even a modest effort might produce a modest harvest, and that it would be less likely that candidates would be

shut out altogether from receiving representation on a state delegation. Whether in response to these incentives or for some other reason, as Table 2.4 indicates, the number of candidates contesting primary elections—especially in the party where no incumbent President was trying to succeed himself—grew sharply after the reforms of 1968–72.[18]

Table 2.4 Number of candidates running in presidential primaries in more than one state, 1952–80

	Republicans	Democrats
1952	4	3
1956	2*	3
1960	3	5
1964	4	2*
1968	3	3
1972	3*	13
1976	2*	13
1980	9	5*

Mean 1952–68 (pre-rules changes): Rep. 3.2, Dem. 3.2
Mean 1972–80 (post-rules changes): Rep. 4.7, Dem. 10.3

*Party of an incumbent President running for renomination

Sources: Robert A. Diamond (ed.), *Presidential Elections Since 1789* (Washington, D.C.: Congressional Quarterly, Inc., 1975); *Congressional Quarterly Weekly Report* 34, nos. 9–24 (1976); Jonathan Moore (ed.), *The Campaign for President* (Cambridge, Mass.: Ballinger Publishing, 1981), pp. 280–88.

(5) With candidate initiatives so thoroughly dominating the delegate selection process, an incentive was created for state parties to separate their internal affairs from the presidential nominating process, thus removing a further impediment to the use of primaries by state parties. A state convention could do all sorts of business. If such a convention were overrun with the zealous supporters of various presidential candidates, politicians with concerns lying closer to home might find themselves inconveniently outnumbered. This created organizational incentives for party regulars—not just party reformers—to switch to primary elections. As Mark Siegel observed preparations for 1972 in New York state:

. . . the state organization was subject to strong criticism [in 1968] for the manner in which . . . non-primary delegate slots were filled. It seemed that the majority of the "regulars" at the guideline adoption meeting, including the organization's leadership, wanted no part of a repetition of the 1968 experience. Indeed, one member of the committee said, "they want a primary, let them have a primary, everything by primary; we don't need more headaches." Thus the committee endorsed all the guidelines and opted for primary election of all delegates and alternates.[19]

Former McGovern-Fraser Commissioner Austin Ranney said, in summary:

Most of the fourteen states which adopted presidential primaries after 1968 did so as a direct response to the McGovern-Fraser rules. Some decided that primaries were the best way to provide genuine "full, meaningful and timely participation." Others decided that the best way to keep the new national delegate-selection rules from upsetting their accustomed and preferred ways of doing state and local party business would be to establish a presidential primary and thereby split off presidential nomination matters from all other party affairs. And still others calculated that the new rules made caucuses and conventions much more vulnerable than a primary to being captured by small but dedicated bands of ideologues.[20]

This bundle of new incentives produced a genuine transformation of the presidential nominating process.[21] From a system in which primaries played a supporting rather than a leading role the United States rapidly moved toward a nominating system in which primaries dominated the process. In 1968, 49 percent of the delegates to the Democratic and 45 percent of the delegates to the Republican National Convention were chosen in states with primary elections, and 36 percent of all the delegates to each convention were committed to candidates by primaries. By 1972 the figures were 66 percent selected and 58 percent committed for the Democrats, and 53 percent selected and 41 percent committed for the Republicans; as Table 2.5 shows, the numbers have increased since then.

We shall be exploring the strategic implications of this set of

Table 2.5 Proportion of delegates selected in primaries, 1952–1980

DEMOCRATS

Year	Number of primaries	Convention total	Delegates selected in primaries		Delegates committed to candidates by primaries	
			Number	Percent	Number	Percent
1952	17	1230	570	.46	224	.18
1956	21	1372	690	.50	523	.38
1960	17	1521	686	.45	311	.20
1964	18	2316	1177	.51	943	.41
1968	17	2623	1276	.49	936	.36
1972	23	3016	1977	.66	1737	.58
1976	29	3008	2264	.75	1982	.66
1980	31	3331	2378	.71	2366	.71

Mean 1952–1968: Selected 48.2%; Committed 30.6%
Mean 1972–1980: Selected 70.7%; Committed 65.0%

REPUBLICANS

Year	Number of primaries	Convention total	Delegates selected in primaries		Delegates committed to candidates by primaries	
			Number	Percent	Number	Percent
1952	15	1206	558	.46	290	.24
1956	18	1323	628	.47	572	.43
1960	15	1331	550	.41	468	.35
1964	17	1308	623	.48	462	.35
1968	16	1333	601	.45	480	.36
1972	22	1348	715	.53	555	.41
1976	28	2259	1501	.66	1219	.54
1980	33	1993	1502	.75	1374	.69

Mean 1952–1968: Selected 45.4%; Committed 34.6%
Mean 1972–1980: Selected 64.6%; Committed 54.7%

Source: Compiled from summaries in *Congressional Quarterly Weekly Reports*.

changes for the workings of the party system and major actors within that system for the rest of this chapter.

2. Factions and Coalitions: Changes in Candidate Strategies

As the Democrats learned in 1972, candidates must behave differently in a presidential nominating process dominated by primary elections than in one in which primaries play a smaller part. Rather than build coalitions, they must mobilize factions.

A political faction is easy enough to define: it is a group acting through a political party in pursuit of a common interest. Factions may give to political parties the loyalties of voters, which contributes to the party's mass base, ideological justifications for party programs, and organizational linkages between leaders and followers. When factions do not operate directly through political parties they are described as "interest groups" or "pressure groups."[22] Whether they are organized and act outside parties or within them, factions, by providing a political focus for the expression of the perceived needs of citizens are by virtue of that fact fundamental entities in any complex political system.

Coalitions are less fundamental structures than factions, in that they are alliances among groups organized for the purpose of achieving goals common to their constituent parts. They arise not out of the natural bedrock interests of people but rather out of their capacity to calculate their advantage over a protracted period, and their ability to see their best interests in light of the complexity of the political world in which they exist. Coalitions come into existence in political systems where two conditions are met: first, no faction is large enough to get its preferences directly enacted by the government without seeking help; second, the system operates by some sort of stable rules giving incentives for factions to undertake strategic behavior. One set of rules common in democratic regimes is majoritarian, in which the assent of majorities or their representatives is necessary for governmental policy to be made. Thus the type and character of coalitions, their durability and their strategies, are determined not only by the sorts of demands made on government, and by the size and strength of the factions that arise to make these demands, but also by the constitutional organization of the polity, with its prescriptions that provide incentives for factions to make alliances with one another.

James Madison claimed that the Constitution he and his colleagues had written was a means for abating the mischiefs of faction.[23] Of course he was right: that is what all constitutions do that provide for institutions meant to be responsive to majorities or their representatives. The means for accomplishing that end are the construction by constitutional provision of institutions that force factions to behave strategically, thus forming coalitions. In the American case there are a number of important coalition-forcing institutions: single-member congressional districts, for example, the bicameral Congress, the electoral college. It is a peculiar feature of a political system so constituted that the task of making and tending the coalitions which feed their preferences through these institutions is itself performed by an extra-constitutional entity, namely, the political party. American political parties organize Congress and almost all the state legislatures, bring voters to the polls, and make nominations for the Presidency. Where party organizations are strong, coalition-building flourishes; where they are weak, the politics of factional rivalry prevails.

This can be seen vividly at the state level. Primary elections have traditionally played the role in state politics of weakening the grasp of political parties. Where any faction can run its candidate for governor, and where many factions frequently do, the capture of state government is determined as if by lottery: as each new candidate enters the gubernatorial race, it reduces the number of votes it takes to vault into the run-off. Thus, the more candidates there are in any given race, the greater the incentive for the next entrant to enter. In states where gubernatorial elections are run in this fashion, factional warfare is frequently at fever pitch. When the incentives are weak for politicians of divergent views to work together, they rarely overcome their natural antipathies. Thus the poisonous political strife of such places as Texas.[24] Meanwhile, even in such an unlikely place as Connecticut—with its deep-seated ethnic rivalries—balanced tickets and mutual accommodations are the norm, owing to nomination processes that make heavy use of such consensus-forcing institutions as state party nominating conventions and the challenge primary.[25]

In the race for the Presidency, the same logic obtains. One of the hidden virtues of the electoral college system—with its aggregation of votes state by state, and the custom of casting all the electoral votes of a state for the winner within the state—is that coalition-building within each state, looking toward the assembly of a majority vote, is encouraged. Under a direct election system, incentives would run in the opposite direction. Instead of pressing politicians toward coalitions that would assemble a majority in each state, factionalism would be encouraged. Each new entrant into the presidential race, carrying with him an unknown potential for capturing the loosely attached votes of a front runner, would create an invitation for the next presidential hopeful and the next until, in a crowded electoral field, differentiation—however marginal—and not consensus-building would emerge as the best strategy for victory. This scenario is, of course, hypothetical for general elections so long as the electoral college persists in something like its present form. There is nothing hypothetical about the impact of its logic on the nomination process, however.

Why must a presidential candidate in the new circumstances created by the proliferation of primaries mobilize his faction rather than build coalitions? The task of a presidential hopeful, threading a path through the minefield of successive primary elections, is not to win a majority but rather to survive. Survival means gaining as high as possible a rank among the candidates running for election. Coming in first in early primaries means achieving the visibility that ensures that a candidate will be taken seriously by the news media. This makes it a little easier to raise money and therefore to contest the next primary. All this hangs on the candidate's being able to attract, let us say, 29 percent of the vote in an early state primary rather than 24 percent of the vote. In a state the size of New Hampshire, in a Democratic primary the difference in absolute numbers of voters between those two percentages may be fewer than 5,000 votes.[26] The candidate's best strategy is therefore to differentiate himself from the others in the race and persuade more of his supporters to come out and vote. A premium is placed on building a

personal organization, state by state, and in hoping that the
field becomes crowded with rivals who cluster at some other
part of the ideological spectrum, or who for some other reason
manage to divide up into too-small pieces the natural constituencies that exist in the primary electorate.[27]

It may not be entirely clear at first blush why a system dominated by primaries exhibits these characteristics more than one
dominated by caucuses. Devoted observers of the contemporary nomination process note that the early Iowa caucus presents much the same survival problem to candidates as the early
New Hampshire primary. It is reported every bit as thoroughly
by the news media and it provides much the same sort of momentum to winners and handicap to losers. Indeed Senator
McGovern in 1972 did unexpectedly well not just in primaries
but also in caucus states, as Norman Miller described in the
Wall Street Journal:

> The reforms created conditions made to order for the well-
> organized McGovern campaign with its legions of zealous volun-
> teers. District-by-district elections of delegates gave the McGovern
> people an incentive to organize early and effectively, even in states
> that didn't look like good bets at first. When the Muskie campaign
> sank, the McGovern forces were already positioned in such states
> as Massachusetts, Pennsylvania and Ohio to exploit the opportu-
> nity.
>
> The McGovern organization's effective exploitation of the reform
> rules has been most striking in little-noticed contests in the 28
> non-primary states. Time and again, bands of McGovern backers
> have "blitzed" precinct-level caucuses and seized power to name
> delegates from outnumbered regulars, who have either backed
> another candidate or wanted to remain uncommitted. The tactics
> have insured Sen. McGovern delegates from a number of unlikely
> states—Texas, Virginia and Oklahoma, to name a few—enough
> that it now seems likely he'll attain his goal of 300 delegates from
> the non-primary states.
>
> Not even organized labor has been able to deliver reliably under
> the new rules. In many places, observes one political operative
> close to the unions, labor has "let a bunch of long hairs and col-
> lege kids beat them." He adds glumly, "The new rules ruined the
> unions—absolutely ruined them."

> While this is obviously an overstatement, it is clear that the tradi-
> tional convention power brokers—labor leaders, governors and
> other political figures—will have their power diminished this year.
> For the reform commission also took action to reduce the number
> of delegates who can be relied on to take orders from bosses.[28]

The reforms not only facilitated the conversion of the nomination process into a system dominated by primaries. They also
vastly increased the role of the news media in the overall process by requiring that delegate selection be done under conditions of great public visibility and in the year of the election.
Thus the early Iowa caucus became a significant early media
event, as Senator Howard Baker described it in 1980, "the
functional equivalent of a primary."[29] Morris Udall discovered
the same thing in 1976, and at the last minute changed his
plans to stay out of Iowa. It was not because the number of
delegates being selected in Iowa were irresistible. Rather, as F.
Christopher Arterton reports:

> Their discovery that the media planned to cover the Iowa cau-
> cuses as extensively as they would the early primaries led Udall's
> advisors to conclude that they could not let the other candidates
> . . . get the jump on them.[30]

Earliness is not the only attribute of a delegate selection process that attracts media coverage. At least three criteria readily
come to mind that might predispose a news organization to give
coverage to a series of delegate-selecting events: the number of
delegates at stake, the competitiveness of the race, and uncertainty about the outcome. Certainly the latter two characteristics are present to a far greater degree early than late in the
election year, and while the careful systematic study of 1980
campaign news coverage by Michael Robinson and Margaret
Sheehan shows some media attention to large states, it is clear
that early states whether large or small, and pre-eminently Iowa
and New Hampshire, dominated news coverage.[31]

As in baseball, to understand the strategic implications of the
selection process, and changes therein, it is important to focus
not only on individual games but on the entire season. What

Table 2.6   News coverage of selection processes, 1980

Date of delegate selection process	State	Number of delegates selected (Dem. and Rep.)	Percentage of total news coverage given over to state primaries and caucuses, 1980 CBS	UPI
Jan. 21	Iowa	87	14%	13%
Feb. 10	Maine	43	4	3
16	Arkansas	52	0[a]	0
17	Puerto Rico	55	1	2
26	New Hampshire	41	14	15
	Minnesota	109	1	0
March 4	Massachusetts	153	7	3
	Vermont	31	1	1
11	Florida	151	3	2
	Alabama	72	1	0
	Alaska	30	0	0
	Georgia	99	0	–[b]
12	Delaware	26	–	0
18	Illinois	271	10	7
25	New York	405	7	6
	Connecticut	89	2	2
April 1	Wisconsin	109	6	5
	Kansas	69	0	1
5	Louisiana	82	0	0
12	South Carolina	62	2	3
	Arizona	57	–	1
17	North Dakota	31	0	0
19	Mississippi	54	0	–
22	Pennsylvania	268	9	7
May 3	Texas	232	2	2
6	D.C.	33	1	2
	Indiana	134	1	1
	North Carolina	109	–	1
	Tennessee	87	–	0
13	Nebraska	49	0	1
	Maryland	89	0	2
17	Dems. Abroad	4	–	0
	Virginia	115	–	0
20	Michigan	223	1	7
	Oregon	78	1	1

May	27	Idaho	38	0	0
		Kentucky	77	0	1
		Nevada	29	–	1
	31	Oklahoma	76	–	1
June	3	California	474	6	4
		Ohio	238	4	2
		New Jersey	179	1	0
		New Mexico	42	0	0
		South Dakota	41	0	0
		Montana	39	–	0
	4	Colorado	71	0	–
				100%	100%

[a] 0 is coverage less than 1% of the total.
[b] A dash denotes that data were not reported by Robinson and Sheehan.

Source: Michael Robinson and Margaret Sheehan, *Over the Wire and on TV* (New York: Russell Sage, forthcoming).

matters is not the type of process in any individual state in any given year but rather the fact that an overall nomination process dominated by primary elections and their electorates tends to force an early definitive choice on the party. Caucus domination of the overall process, on the other hand, holds open some options for state party leaders. They may be able to remain uncommitted and later on can make bargains once they inform themselves about the range of alternatives available to them. Primary electorates almost always prefer to cast their ballots for a well-advertised somebody rather than the nobody entailed in a favorite son or uncommitted vote.[32] A nomination process consisting, as it now does, mostly of primary electorates and the registration of their preferences effectively kills an otherwise live option to stay uncommitted that might exist in the caucus states. Pressure on state party leaders from candidates and their in-state allies is very great. The glare of publicity is hard to resist. The rules favor committed delegates, even at the lowest levels. And so caucus states in nomination systems dominated by primaries exhibit most of the characteristics of primary states, and the strategies that must be used in dealing with the caucus states are strategies dictated by the overwhelming presence of primaries in the process.

3. *Changes in the Influence of Party Elites*

The mounting of a presidential campaign has come more and more to resemble the production of a Broadway show. A company is newly created. It sells tickets, books theaters, writes a script (frequently known as "the speech"), and advertises the star. Individual contributions are solicited through the mail. As these build up, candidates become eligible for federal matching funds provided by law to assist candidates successful in their initial fund-raising efforts. With these funds, they can advertise themselves further and campaign in additional state primaries.

Rather than depending upon alliances with and commitments from state party organizations or interest groups allied to factions within state party organizations, candidates for the Presidency are increasingly obliged to mount their search for delegates by building their own personal organizations, state by state. This is necessary because frequently state party organizations wish to maintain neutrality toward all primary contestants in the hope of developing a decent working relationship with—or at the least of not alienating—the eventual winner of the nomination. State party organizations also may find it difficult to deliver the vote for the candidates of their choice in primary elections in which opposing candidates are effectively advertised or for other reasons enjoy high name recognition. Consequently, candidates must themselves advertise and promote high name-recognition as a means not only of winning delegates but also of enlisting, or at least neutralizing, state party leaders. These incentives for candidates can arise in the following way:

(1) The proliferation of primaries weakens the influence of state and local politicians on the choice of delegates and increases the influence of the news media. Free and favorable publicity in the news media is bound to be more directly helpful in influencing a primary electorate than in influencing a party leader.

(2) Favorable publicity aids in the collection of funds when federal subsidies are conditioned on broadly based fund-

raising, mostly by mail solicitation. Anonymous givers by mail are far more likely to send money to candidates of whom they have heard.

(3) Money helps candidates contest primary elections, which are very expensive because of the need to use the mass media. Newspaper and television advertising and television production and all the other modern devices for reaching mass electorates demand far more money than whatever it used to take to reach state party leaders and prospective delegates.

Developments like these have given rise to the displacement of state party leaders and leaders of interest groups associated with them in the presidential nomination process—the demise, so to speak, of presidential nominations as repertory theater. It does not mean that candidates shun all party leaders or factional leaders within the states, but rather that selected state leaders may be invited to participate in a candidate's campaign rather than the candidate being a recruit and representative of the state party or faction. The balance of initiatives, and of responsibilities for the overall organization of the effort, has definitively shifted. It has also meant that a new group of political decision-makers has gained significant authority. These are the fund-raisers by mail and by rock concert, media buyers, advertising experts, public relations specialists, poll analysts, television spot producers, accountants and lawyers who contract themselves out to become temporary—or in the case of a successful candidate, perhaps longer than temporary—members of the entourages of presidential aspirants. They work not for the party but directly for candidates.

The growth of campaign management as a profession demanding the mastery of increasingly arcane and complex bodies of knowledge was noted early and has been carefully monitored by students of presidential nominating politics.[33] In two respects political reform has accelerated this process of professionalization.

First and most important, technocrats in the candidates' entourages have enjoyed enormous increases in their influence because of the decline in the power of party politicians. Second, the techniques that technocrats master have become more complicated and more demanding. The Federal Election Commission, created by the 1974 reforms of campaign finance, now issues streams of memoranda regulating the receipt, reporting, and expenditure of campaign funds.[34] Knowing what needs to be complied with is itself a sizable challenge, not to be undertaken without the assistance of a lawyer and an accountant.[35] Candidates used to find out all they thought they needed to know about public opinion by talking to other politicians, or by reading the newspaper, or by taking the temperature of the crowds to which they spoke.[36] Nowadays, professionally drawn and certified public opinion surveys have largely supplanted seat-of-the-pants methods.

Owing to the stern prohibitions of the federal law, financial angels now fear to tread in the precincts of the candidate. Rushing in to take their place are the professionals who run mail solicitation operations, impresarios who can put together and deliver authentic media stars to fund-raising events, and increasingly, the access-maximizing professional agents of Political Action Committees, who invest in candidates in behalf of groups of contributors. Only the last of these are employed independently of the candidate's entourage.[37]

Commercial messages broadcast on radio or television touting the merits of candidates have never been in the hands of amateurs. The sophistication of the professionals who now do this sort of work is nevertheless easily visible to those who remember the election night broadcasts of 1952 and 1956 and can compare them with the sort of thing broadcast today, just as today's network evening news program bears only a faint resemblance to the old Camel News Caravan.

The great issue of democratic theory that arises in connection with the decline of one set of political elites and the rise of another is of course the issue of accountability, as we will see later in tracing through the implications of party change for interest groups. For now it is enough to ask: what values animate and constrain the behavior of these new political elites? How do these compare with whatever the values were that animated and constrained the old?

In some cases the answer is professional values. Lawyers, accountants, and even managers of public opinion polls have standard ways of doing business that as a matter of professional ethics they require their colleagues to observe. A polltaker who knowingly divulges to his client the name of a respondent who has been offered anonymity, or an accountant who claims to have audited an account when he has not done so, risks severe professional sanctions.

There is also the sanction of the law: accepting or spending more money than the law will allow is not a professional error in accountancy—it is a crime. The existence of far more regulations in the political environment, the violation of which lead to criminal charges is, however, less a constraint upon new political elites than an opportunity for them, since this creates a situation in which candidates feel—and may indeed be—unsafe without professional help.

Whatever the professional constraints on the behavior of technicians of political campaigns, however, it cannot be said that they represent anyone. The access they gain to candidates serves no legitimate interest beyond that of their own careers, and presumably the careers of the candidates whom they serve. The peopling of a campaign organization with specialists thus makes no coalitions, cements no alliances, seals no deals in the world of grass roots electoral politics. In this sense the professionalization of campaigns has served to isolate candidates even though, in another sense, they may be better informed than ever before.

4. National Party Conventions

It follows that the institution of the national party convention, as everyone has noticed, has been transformed. Instead of a body of delegates from the state parties meeting to ratify the results of a complex series of negotiations conducted by party leaders at the convention, the convention is now a body dominated by candidate enthusiasts and interest group delegates who meet to ratify a choice made prior to the convention mostly through primary elections. A major part of the responsibility for this change can be assigned to the effects of television, which has the power to start bandwagons very early in the delegate selection process. Reforms have not been intended to increase the influence of television: abolition of the unit rule, for example, was in part a device to spread the rewards for candidate effort and to prevent the premature consolidation of delegates by an early winner. On the whole, however, the enhancement of incentives for candidate initiative has had as its dominant consequence the weakening of state party leaders, and this has in turn given to primary voters and contributors to mail financial solicitations—both populations overwhelmingly dependent for their information about politics on television and other forms of mass publicity—the major role in the pre-convention nomination process. Thus the publicity effects of television have overridden the efforts of party leaders to tinker with party rules to contain their influence.[38]

So bandwagons have not merely survived the changes in the presidential selection process; they have flourished. They operate earlier and earlier in the process. Indeed, everything consequential occurs earlier than ever before. Candidate initiatives must begin earlier as well, indeed so much earlier that it is now accounted an advantage for a serious presidential candidate to be unemployed or at least without major distracting responsibilities for a full year or more preceding the nomination.[39]

Delegates are for the most part pledged to a candidate before setting foot in the convention hall.[40] Much ink has been spilled worrying over the legal enforceability of such a pledge, made in one state according to state primary law but to be redeemed in another state, the locus of the national convention, in a body controlled by its own rules and backed by federal law.

The 1980 Democratic convention was briefly amused by a preliminary skirmish—predominantly a thinly disguised attempt to prevent the nomination of Jimmy Carter—over a proposed procedural rule making the honoring of a delegate's pledge enforceable by the convention itself and by the candidates.[41] Without such a rule, it was argued, delegates could exercise their obligation freely to deliberate on the business of the convention. As a practical matter, delegates pledged to candidates in each state are cleared if not actually named by the candidates and their agents. Such delegates represent nobody but the candidate to whom they are pledged. It is difficult to see in whose behalf they might presume to deliberate in the unlikely event they were to decide to do so. Delegates pledged to a candidate and elected on a candidate's slate in a primary have no standing to deliberate except as that candidate's representative even if called upon to make the attempt.

Since the transformation of 1968–72, of course, no deliberation over the nomination has taken place at conventions. Dark horses—candidates who arrive at the conventions with fewer delegates than the front runner and who hope to pick up delegates as a plausible second choice when more favored candidates cancel one another out—have become an endangered species. Indeed all strategies based on second choice strength have disappeared. Such strategies entail coalition-building, not factional mobilization.

So today national conventions survive primarily as spectacle—an ingathering of the multitudes who by their good behavior can reward their foreordained nominee with favorable publicity, or by their bad behavior can cripple the party's ensuing presidential campaign. More and more conventions are designed as entertainment, although interstitially they may still conduct business of great importance to the future of the party, as, for example, when they pass resolutions governing the future conduct of party affairs, such as the Democratic resolution in 1980 forbidding national committee funds to be used in support of party candidates not in favor of the Equal Rights

Amendment[42] or the Democratic resolutions in 1968 establishing a commission on delegate selection.

5. *Central Control vs. Voluntarism*

Centralization within each national party has resulted as a consequence of new rules. Each national convention under the law has firm control of its own membership, and can set criteria which state parties must meet in order to be seated at conventions. Broad party principles require means of enforcement and particular application, which emanate from the national committees. Well in advance of national conventions compliance review must take place, so that state party adherence to criteria laid down by the previous national convention or by other nationally sanctioned instrumentalities can be assured. Immediately prior to national conventions this review can be undertaken by credentials committees run by delegates themselves.

Before the convention is constituted, however (bearing in mind that delegates may not be selected too far ahead of time without violating rules requiring timely delegate selection),[43] consultation, persuasion, and informal adjudication on state party delegate selection procedures and activities are necessary. The differential between the major parties on this matter has been sharp, with the Republicans having few centrally mandated criteria to enforce, and little appetite for centralized enforcement activity. Among the Democrats, the exercise of rule-application functions has to all intents and purposes become a prerogative of national committee employees. Not all these employees have in the past been perceived by state party leaders as embodying the soul of tact; in some cases senior, or at least duly elected leaders of state parties have taken unkindly to jawboning from zealous young agents of a formerly moribund and—despite the best literary efforts of the Supreme Court[44]— still only questionably legitimate central party bureaucracy.

The sense of loss of autonomy among state party leaders in national affairs has conceivably turned their concerns inward toward state level concerns and reduced their incentives to contest with various interest groups and candidate enthusiasts for influence over the presidential nomination process. Thus centralization of power through the compliance review process in the hands of national party functionaries can lead to a weakening of party organization at the intermediate level of the state parties. Whether this in turn permits the grass roots to flourish is a matter for consideration in a later chapter.

A further consequence of the new atmosphere of regulation, professionalization, legal responsibility, and legal sanctions is the centralization of political activity within each candidate's campaign, a logical concomitant of the centralization of political authority and of criminal liability. Richard Cheney, who ran Gerald Ford's campaign in 1976, testifies eloquently on this point:

> One of the major results of the spending limitations has been to encourage the development of highly centralized campaign organizations with elaborate controls over spending. Unless a campaign develops such an organization, there is virtually no possibility that it can account for all of the funds expended or adequately comply with federal regulations. While this makes for a more efficient campaign operation, it has had the effect of choking off the kind of grass-roots activity that historically has been a part of American presidential campaigns.

> The experience of the Ford campaign in 1976 showed conclusively that it was easier to discourage grass-roots activity than to try to control and report it. In previous campaigns, it was possible to tell a local campaign or party official to go ahead with a project as long as he could raise the money to finance it. Now, federal law places a premium on actively discouraging such activity because of the danger that it could well lead to a violation of contribution or spending limits in the primary. Furthermore, in the general election, because no contributions are permitted once federal funds become available, it is even more important to discourage such activity.

> Such considerations lead to shifts in spending priorities and, therefore, campaign strategies. State-by-state primary spending limitations, the overall limitation on prenomination spending, and the requirement that none of the money raised before the con-

vention be used to promote the general election effort (which has to be totally financed with federal funds), all serve to discourage organizational activities. We found it much easier during the 1976 Ford campaign, for example, to spend money on identifiable goods and services, such as electronic media and production costs, in the general election campaign than to spend it intelligently on local and state organizational efforts. This was especially true because we had discouraged organizational activity that was not directly under our control during the primaries. The case of the Florida primary, held in March 1976, is illustrative of the difficulties posed by the new law. The Ford campaign poured significant resources into the Florida effort and won our second major victory over Governor Reagan. Within days of the election, the entire Florida operation had to be totally shut down because of the various limitations we faced. As a result, there were no resources available to keep even a symbolic operation going through the summer in preparation for the fall campaign—no headquarters facility, no phones, no paper clips, and no staff.

The same thing happened in virtually every state where we contested a primary in 1976. It was hardly surprising, then, that with only a little more than two months for the general election campaign, we found it difficult to spend money on organizational efforts at the state and local level when we had dismantled the nucleus of our organization at the end of each primary campaign. It made a lot more sense to spend it on media.

I firmly believe that the effect of the campaign finance laws in this area has been to discourage grass-roots political activity, to discourage participation, and to place a premium on strategies that rely on activities that are easily controlled and reported. Given a choice between local spontaneity and enthusiastic participation, on one hand, and control over spending, on the other, the cautious campaign manager has little choice but to opt for activities that are "controllable."[45]

The net effect on state and local party organizations of this discouragement of grass roots political activity during a presidential election year is hard to assess. At a minimum, it tends to diminish the value of state and local party leaders in their dealings with candidates, thus reinforcing trends that already have ample momentum. Further, on the assumption that presidential campaigns are a good way to build party solidarity at the grass roots, to recruit workers and to reinforce their loyalties and their commitment to the party as an organization, a consequence of political reform might well have been to encourage the atrophy of political parties at the grass roots.

Niggling restrictions of the Federal Election Commission have contributed to the drying up of voluntarism: campaign managers have had to worry, for example, whether they were liable for an assessment against the total expenditure permitted to a presidential candidate for the cost of lapel buttons bought and paid for by a local congressional candidate's campaign because the names of both the congressional and the presidential candidate appeared on it.[46] Casual observation detects the decline of the most decentralized of campaign aids: buttons, bumper stickers, and billboards. Expenditures that directly benefit the top of the ticket—television appearances, moving the candidate around by aircraft—continue to soak up lavish amounts of money, however. But cooperative expenditures, where central fund raising is used to nourish the grass roots of the party, have shriveled.

So serious was the problem thought to be that amendments to some of the restrictions on campaign expenditures were brought forward in 1979. Congress removed prohibitions against the purchase of buttons and bumper stickers for volunteer activities by state and local parties, and allowed incidental mention of presidential candidates in other candidates' literature without the mention being considered a contribution. Spending limits on volunteer activities were eased, and get-out-the-vote and some kinds of voter registration drives were allowed to party organizations without financial limit. It remained to be seen, however, whether these changes would be enough to overcome all the obstacles to local effort that had been erected by the new rules.[47]

6. Consequences for Party Competition: Third Parties

Thus far we have considered consequences of party reform for the two major political parties as organizations, from the standpoint of their internal functioning as machines for making presidential nominations and conducting election campaigns. Party reform has in addition had an impact on the party system, that is, on the relations between parties and on the ways in which they compete for public office. Presently I shall deal with the impact of reform on competition between the two major parties, but here I wish to consider briefly the effects of party reform on third parties.

There are at least three sorts of disabilities that stand in the way of the success of third parties in American national politics.[48] One is ideological. Because of the doctrinal looseness of the two major American parties, their continental reach into widely differing milieus, and their nonexistent formal criteria for membership, it is hard for a third party to get much of a toehold on ideological grounds. All popular ideologies are in some form or other generally embraced by the two major parties. The genuinely disaffected in a political system like the one prevailing in the United States are far more likely to be utterly withdrawn and unavailable for political participation than they are to be eagerly clustered in some convenient spot on the map—geographic or social—waiting for a new banner to be raised. Those inclined to participate can frequently do so on ideological terms close enough to their own to make third parties no more intrinsically attractive than one or another of the first two. By world standards the Democrats and Republicans may not cover much of the ideological spectrum, but they seem to serve their markets uncommonly well.[49]

Third, fourth, and fifth parties have from time to time emerged, and from place to place have flourished as, for example in the cases of the Liberals and Conservatives of New York. On the whole, however, minor parties have made little headway among voters. At any given time in the mid-twentieth century only a handful of officeholders at any level of government in the United States belongs to some other party than the Democratic or the Republican parties.[50]

In the second place, the bipartisan condition is facilitated by the constitutional forms of American politics: single-member districts and plurality elections, sometimes with run-off provisions, create their well-known premium on coming in first. The electoral college, likewise, operates at the presidential level to encourage the building of a winning coalition broadly based among the states, and by depriving those who lose narrowly in a state from sharing any of that state's electoral votes it discourages weak candidates and penalizes third parties.[51]

Finally, there are many and varied discouragements to third parties written into election laws and regulations having the force of law. Ironically, many of these disabilities are the consequence of an intention to foster party competition. In most cases if they encourage competition at all, it is distinctly two-party competition. These problems are not new, for, as an example, some state statutes have long constituted election boards only from members of the two major parties.

Recent reform activity has nevertheless increased the burdens upon third parties and independent candidates. A massive example is provided by the Federal Election Campaign Act Amendments of 1974 and later years, which provide public financing for the campaigns of the two major political parties but delay payment to new parties until it is determined—after the election—that they got enough votes to qualify as nonfrivolous and may exclude independent candidates altogether, while subjecting them all to the same limitations in their receipt of funding from private sources.[52] Major parties are allocated public funds for holding their nominating conventions; minor parties are not.[53]

Congress has likewise frequently suspended equal time and fairness doctrine requirements on F.C.C. broadcast licenses so as to make debates between major candidates possible—and to make possible the exclusion of minor party candidates and independents.[54] Minor parties are also disproportionately burdened by the strict disclosure requirements of the new federal election law with respect to financial contributions, since they have reason to believe that prospective contributors may be shy of being publicly identified with political candidates who are out of the mainstream and who may, indeed, be widely considered to be odd. This is sometimes known as a "chilling effect."[55]

There is a practical way out of a dilemma which confronts the presumed chilling effect on third party contributors who might fear social stigmatization or worse with the presumed civic advantages that follow from public disclosure—and prior knowledge that public disclosure will take place—of contributors to and financial backers of the candidates who will almost certainly win the election. And that is to offer small deviant parties exemptions from reporting requirements, as courts have been asked to do in the case of the Communist and Socialist Worker parties of New York and Pennsylvania, respectively, and that of a self-identified homosexual candidate running for county commissioner in Ohio. There may, however, be no other practical way to hold a televised debate between the two candidates who monopolize the real chance of being the next President than arbitrarily excluding the rest. The reinforcement effect is important, however. When television networks exclude candidates from a debate or a joint appearance with the front runners they are denying them visibility, publicity and, most of all, a social definition as being worthy of serious consideration. An attempt to find a formula based on the standing of a candidate in public opinion surveys, as was tried in 1980 with John Anderson, suffers from a possible confusion between cause and effect: the easiest way to ensure that Anderson met the threshold number would have been to put him on television, visibly equal to the major party candidates. His exclusion, conversely, was the best guarantee that he would diminish in public esteem.[56]

Much the same problem arises when we consider public finance. Some rational basis for the expenditure of public funds has to be established if candidates are to be publicly funded. Otherwise, incentives are created for frivolous candidacies. Limitations on subsidy, however, are unavoidably limitations upon political expression and have the effect of entrenching the two major political parties.

7. Competition Between the Major Parties

Perhaps the most startling effect of political reform has been its influence on the competitive balance between the two major parties. The smaller, more ideologically cohesive of the two, the Republican, has reaped enormous benefit from changes weakening the internal organizational integrity of both parties.[57] Where ideology must substitute for institutional rules in maintaining organizational cohesion, a party with a strong ideological core and not much periphery may be better off at least over the short run than a party like the Democratic that attempts to embrace a wider range of ideological perspectives.

It may not be immediately apparent why political reforms established across the board might affect the two major parties somewhat differently. In some respects, to be sure, they do not. Although most of the dramatic examples of changed circumstances come from the Democratic side, Ronald Reagan's almost-successful challenge to President Ford in 1976 would have been far less likely under a system with fewer primary elections. Nevertheless, the two major parties are somewhat different in their organizational, coalitional, and ideological makeup, and the impact of the reforms has differed accordingly.

The key difference is ideological. Factionalism in the Democratic party, a failure to build a broad coalition, exposes Democratic candidates in the general election to the full force of the wide variation that Democrats of different stripes maintain in their policy concerns and commitments. Failure to embrace a coalition-building strategy risks defection in such circumstances far more surely than in the Republican case, where the party maintains a high rather than a broad church of co-believers and where, consequently, serious contestants for the Republican nomination are much more likely to think more or less alike about public policy. The classic Republican problem in presidential elections has not been the shaky loyalty of Republican voters but rather the party's too narrow base in the general electorate. Thus, Republicans internally have less to fear from a nomination process that encourages factionalism and discourages coalition building since Republicans are already more nearly factionally cohesive and like-minded on political issues.

Centralization has also posed less of a problem for the organizational side of the Republican party, because central management of an organization with a narrow ideological dispersion means that party managers will be making fewer choices fraught with political implications as far as party members are concerned, and a broader range of functions will be considered to be purely ministerial in character. Moreover, Republicans have a lot more money to devote to party management. The central offices of the Republican National Committee have in general been better staffed and its business more stably and professionally conducted than has been true of the Democrats.[58] Under the innovative management of William Brock, professional services were expanded for identifying, recruiting, training, and supporting Republican candidates for elective office at state and local levels.[59] But party rules have forbidden meddling: in 1980 the Convention adopted a rule requiring state party consent before the Republican National Committee could contribute from its bulging war chest to candidates within a state.

A stunning consequence of the differential impact of reform on the two major parties is that in two out of the last three presidential elections the minority party has won and the majority party has lost.[60] Thus, it is at least arguable that party reforms have worked disproportionately to the disadvantage of the majority party, the Democrats, by leading to the nominations of candidates unable to command widespread support within the party. The reforms in effect have institutionalized the disequilibrium suffered by the party in 1968 and facilitated Republican victories in 1972 and 1980. In 1968, as we have seen, the majority party tore itself to tatters. Even so, it took an assassination to destroy their otherwise reasonable chance for victory.[61] The Democratic rules changes of 1968–72 weakened the state parties and so strengthened candidates in the nomination process that in 1972 a candidate widely perceived as an extremist, and certainly supported by extremists,[62] could win the nomination and singlehandedly in the general election ward off a Democratic landslide. While the Democrats won the 1976 election, I shall presently argue that who their candidate was and what he felt able to do in office were importantly shaped by the constraints, or lack thereof, of the nomination process. And of course the majority party candidate in 1980—and the loser of that election—was an incumbent President; hence responsibility for his nomination in that year rests primarily with the workings of the process in 1976. If majorities are supposed to win in a democracy, then the American presidential election process has not lately been mobilizing them.

Some of the underpinnings of this argument can be readily noted. We can certainly begin with the fact, already established, that the Democratic party is without question the larger of the two major parties in terms of long-term self-identified adher-

Table 2.14 Defection levels from party in presidential voting, 1952–80

| Year | Democrats | | | By comparison, |
	To Rep. candidate	To 3d party (if signif.)	Total	Republicans (to anybody)
1952	23%		23%	8%
1956	15		15	4
1960	16		16	5
1964	13		13	20
1968	12	14%	26	14
1972	33		33	5
1976	18		18	9
1980	26	4	30	13

1952–1968 mean defection: 18.6% Democratic; 10% Republican
Post 1968 mean defection: 27% Democratic; 9% Republican

Source: The Gallup Opinion Index Report No. 183, December 1980, pp. 6–7.

ents.[63] Republican Presidents have repeatedly been elected by
the votes of Democratic defectors, who give as their major reasons for voting Republican their dissatisfactions with the Democratic alternative.[64]

Claims that Republican Presidents are elected not by Democrats but by formerly inactive or disaffected or latent Republicans founder on all sorts of difficulties: the fact that at or around
the time of the elections in question heavy majorities rejected
Republican positions on many controversial political issues,[65]
that Democrats retain the nominal loyalties of more voters than
Republicans,[66] that Democratic votes have consistently outnumbered Republican for all other offices.[67]

We can establish as well that the reforms put in place organizational features of party structure that made it difficult for
anyone to do anything effective about forestalling the nomination of candidates who may have been extremely unpopular
with large numbers of party members. This is a clear cost of
the automated, nondeliberative national nominating convention in which state party leaders and officeholders play increasingly passive roles, and where proceedings are dominated by
candidates who have previously secured a majority of pledged
delegates by accumulating sufficient first-place wins in state primaries and conventions to stampede the uncommitted remainder onto their bandwagon.[68]

All this adds up to a plausible, perhaps even compelling case
for the proposition that various reforms of the presidential selection process have had a sizable impact on American political
parties and the party system. To be sure they were intended to
affect the parties, though perhaps not to the extent, or in each
and every way, that they have done. But because the parties
are the principal agency in our political system through which
future Presidents are recruited, it might have been foreseen
that the effects of presidential selection reform would not stop
with the parties, or even the party system, but would also influence the conduct of the Presidency as well.

III

Consequences for Governing: The Conduct of the Presidency

Since the election of 1968, the Democratic majority has been able to pull itself together sufficiently to elect a President only once. This adds up to twelve years of Republican rather than Democratic control of the federal government in the sixteen years between 1968 and 1984, a rather sizable effect to be attributed in more than trivial ways to party reform. I believe it can reasonably be maintained, in addition, that party reform also had an impact on the four intervening Democratic years. For the Democratic incumbent, Jimmy Carter, was a President whose conduct in office to a remarkable extent faithfully reflected the learning experiences available along the pathway he followed in order to achieve the Presidency. The argument in brief is that President Carter conducted himself in office in ways that were fully consonant with his personal predilections and his views of public administration but which would have been harder for him to pursue if he had been educated in the course of the nomination process to the need to build a governing coalition. What it takes to achieve the nomination differs nowadays so sharply from what it takes to govern effectively as to pose a problem that has some generality. This problem can conveniently be illustrated by a closer look at the Carter Presidency and its difficulties in the context of recent Presidents.

1. The Cabinet: Recent Historical Experience [1]

There was, to begin with, the early evidence of the composition of President Carter's cabinet. In general, the pattern of cabinet appointments a President makes can be informative about what sort of Presidency he means to have. Like the campaign rhetoric that precedes the election, a newly appointed cabinet can be read in a variety of ways. And it affords less than comprehensive evidence about how the president plans to run the government. But fragmentary though it is, this sort of information is hard to ignore, for unlike the rhetoric of campaign promises, cabinet members do not disappear into thin air.[2] Rather, they assume office and to a greater or lesser extent actually administer the affairs of the nation. President Eisenhower's appointment of "nine millionaires and a plumber" gave a good forecast of the sort of status-quo-defending Presidency he wanted.[3] It could not have come as much of a surprise when the plumber, Secretary of Labor Martin Durkin, was the first to leave the cabinet. When John F. Kennedy became President he struck a dominant theme of self-consciously moving beyond his own range of personal acquaintance to form a governing coalition. As Douglass Cater said:

> While consulting a senior statesman about the selection of a Secretary of State . . . Kennedy remarked that he had suddenly discovered he didn't know "the right people." During his campaigning he had, of course, met practically every politician in the country. But as far as picking a cabinet was concerned, his large circle of acquaintances seemed inadequate. The truth of these remarks . . . was subsequently borne out when Mr. Kennedy appointed men not previously known to him to several key posts in his administration.[4]

In one conspicuous case, of course, Kennedy was well acquainted with a cabinet appointee: His appointment of his brother as Attorney General telegraphed—among other things—his strong desire to keep close control of the civil rights issue.[5]

It is possible to see in Richard Nixon's cabinet appointments a mirror of his emerging view of the role of the President vis-à-vis the rest of the government. After beginning with a group of cabinet appointees that was both politically diverse and reasonably visible, Mr. Nixon increasingly appointed people with no independent public standing and no constituencies of their own.[6] In this shift we can read a distinctive change in the fundamental political goals and strategies of the Nixon administration from early concerns with constituency building to a later preoccupation, once Mr. Nixon's reelection was assured, with centralizing power in the White House.

Table 3.1 Decline in prior experience of Nixon cabinet

	President Nixon's First Cabinet, 1969	*President Nixon's Last Cabinet, 1974*
Prior political experience extensive (includes office-holding)	William Rogers, State Melvin Laird, Defense Walter Hickel, Interior Maurice Stans, Commerce Robert Finch, HEW George Romney, HUD John Volpe, DOT	Rogers C. B. Morton, Interior Earl Butz, Agriculture
Prior political experience moderate (active in state party, etc.)	Winton Blount, Post Office	Frederick Dent, Commerce Peter Brennan, Labor Caspar Weinberger, HEW
Prior political experience slight	David Kennedy, Treasury John Mitchell, Attorney General Clifford Hardin, Agriculture George Shultz, Labor	Henry Kissinger, State William Simon, Treasury James Schlesinger, Defense Robert Bork, Acting Attorney General James Lynn, HUD Claude Brinegar, DOT

From Nelson W. Polsby, *Congress and the Presidency*, 3rd edition, (Englewood Cliffs, N.J.: Prentice-Hall, 1976), p. 53.

Nixon's first Secretary of Labor, George Shultz, though unknown to begin with, became an early star of the cabinet owing
to his intelligence and quick grasp of problems. The first major
reorganization of the Nixon Presidency shuffled Mr. Shultz into
the White House.[7] He was replaced by an efficient but unpre-
possessing figure (James Hodgson), who in turn gave way to a
maverick union official (Peter Brennan), who was not even on
speaking terms with the head of the AFL/CIO.[8] This was not
the only example of a movement away from clientele concerns
in cabinet building and toward the accretion of managerial capacity within the White House. Seemingly by design the access
of large and significant interest groups to the President was
greatly hampered. Labor, education, the scientific community,
conservationists and others felt not merely that Richard Nixon
was a President whose goals differed from their own but that
their voices were being choked off, that they were shut out from
the White House and that their cases were being rejected before they were heard.[9]

If Mr. Nixon's administrative appointments were designed to
be increasingly weak in their capacity to carry messages from
interest groups to policymakers, they were far stronger in executing orders, in providing a conduit from the various arms of
the White House executive apparatus—the Domestic Council,
the Office of Management and Budget, the National Security
Council—to the levels of policy execution.

Centralization of policy making was only half of the latter
day Nixon administrative program. The other half consisted in
systematic attempts to place functionaries who can best be described as political agents in the bureaus and departments, ad-
ministrators whose job it was to report to the White House on
the political fidelity of the executive branch.[10]

Presidents and their political appointees have frequently
puzzled over the problem of making the enormous apparatus
of the executive branch responsive to their wills. The legitimacy
of this claim is based upon the results of the last election; presumably a President is elected and makes appointments at least

in part to carry out his promises with respect to the future conduct of public policy.[11] The necessary instruments of that conduct reside in the unelected agencies of the executive branch.

Executive agencies, however, are seldom merely passive re-
ceptacles, awaiting the expression of the President's preferences. Rather they embody a number of persistent characteristics that from time to time may serve as bases for conflict with
presidential directives. Expertise, for example, a body of doctrine about the right way to do things, may well thwart responsiveness to presidential demands. So may alliances between
agency executives and the Congressional committees that have
program and budgetary oversight over them. So may strong
ties between agencies and the client groups they serve.

The case of a conservative President facing an executive
agency doing what he believes is liberal work is especially poignant. The very existence of a bureaucratic apparatus attests to
the mobilization at some time in the past of a majority of sufficient strength to pass a law and put an agency to work. So
long as the law is on the books and Congress appropriates funds,
the agency presumably has some sort of legitimate standing.
Yet it was precisely the existence of all too many of these federal activities, all staffed with people devoted to the execution
of their programs, that Mr. Nixon wished to challenge. In one
famous instance, the case of the Office of Economic Opportunity, President Nixon attempted prematurely to put an agency
out of business altogether.[12]

In each of Mr. Nixon's attempts to organize and reorganize
the executive branch, an observer can note efforts to cope with
what was evidently being defined as a hostile administrative environment. Revenue sharing, of course, had the intended effect
of removing responsibilities altogether from federal agencies.[13]
Mr. Nixon vastly strengthened the White House National Security Office and invented a domestic counterpart in the Domestic Council, politicized and reincarnated the Bureau of the
Budget as the Office of Management and Budget, attempted
on a sizable scale to impound and redirect funds appropriated

according to law, and drastically increased the number of employees in the Executive Office of the President.[14] Just as the storm of Watergate was breaking over his head, he proposed a reorganization plan that would have officially denied cabinet officers direct access to the President by shifting supervisory power to four super-cabinet officers who were supposed to act as special presidential assistants.[15]

These devices for limiting the power of government departments, agencies, and bureaus were in general not illicit. They reflect a distinctive view of executive branch legitimacy and its monopoly in the presidential office. Mr. Nixon's view, it became clear, was that his reelection in 1972 by a landslide not only provided him with a special entitlement to pursue his vision of public policy, but it had in addition delegitimized all other major actors in the system.[16]

The last pre-Watergate months of the Nixon Presidency saw neither the first nor necessarily the last official manifestation of the view that the President is the source of all the legitimacy on which the entire government runs. Indeed even today the theory is widely held that, because the President is the only elected official in the executive branch, political choice by the executive branch is legitimized only insofar as it can be plausibly seen to have radiated down from a presidential choice or order or preference.

Sustenance for this view comes from a conception of the American political system in which legitimacy arises chiefly, if not exclusively, from the electoral process. If the direct results of elections are the only source of political legitimacy, then it follows that the legitimacy of the entire executive branch flows as though down the sides of a gigantic pyramid from the source: the quadrennial election of the single member of that branch who is elected—the President. Given this view of the situation, there can be no grounds upon which hierarchical subordinates of the President might legitimately act to thwart, undermine, delay, modify, or attenuate the will of the President in public policy once it is expressed.

This is not, however, the only possible view of political legitimacy in the executive branch. Another view is that even though the electoral process does provide a significant measure of legitimacy for the acts of government, this process by itself is neither adequate to the task of providing accountability, nor is it the only process actually provided for in the constitutional design and in the pattern of American politics that has since evolved in harmony with the spirit of the Constitution. In contrast with this hierarchical or pyramidal or plebiscitary view of legitimacy, there is the check and balance or multiprocess view favored, for example, by the authors of the *Federalist* in which the rightful power to govern is spread about in the government and even, in a more modern version, extended to interest groups and other mobilizers and organizers of popular desires, needs and sentiments.[17]

It is interesting to contemplate the extent to which the Presidency-centered view of the proper relations between the Presidency and bureaucracy survived intact through Watergate and indeed through the fashionable disparagement of the imperial Presidency.[18] This was reflected in President Carter's initial address to the problem of cabinet building and what appeared consequently to be his first perspective on the permanent government of the United States.

2. *Five Strategies of Cabinet Building*

There are at least five different ways to build a cabinet. The first option takes account of the fact that each of the great departments of government serves clientele in the population at large. Each has custody over a range of policies that tend to affect some Americans more directly than others. Thus one strategy for building a cabinet is to enter into a coalition with the client groups of the departments by finding appointees who already have extensive relationships or political alliances or management experience with relevant client groups. A cabinet in which this alternative is dominant is frequently one heavy with former political officeholders.

Characteristic costs of this mode of cabinet building include those associated with complaints from interest groups whose competitors have succeeded where they have not. So, for example, conservationist groups may feel exceedingly well served by the appointment of one sort of Secretary of Interior. Grazers, miners and loggers may feel quite differently about the matter. The impossibility of accommodating all the groups into which Americans may legally divide themselves for the purpose of pursuing a common interest is no doubt one of the facts of life that give vitality at the presidential level to a competitive two-party system. The fact that ungratified client groups can form alliances with the out party helps to legitimize the inevitable choices that Presidents must make among contending interests. This is one mechanism for expressing dissent from governmental policy which, if it becomes loud enough or effective enough, can limit the extent to which any incumbent may persist in a line of policy without risking electoral defeat.

An advantage of this strategy is that by drawing from a pool of politicians associated with clientele of the agencies, presidents may cement electoral alliances with interest groups and draw them into collaboration with his administration, forming a governing coalition that resonates throughout the population at large.

A second option is available to a President who may, if he chooses, select a cabinet of substantive specialists. Specialists possess technical mastery, knowledge of programmatic alternatives and understanding of particular governmental agencies and their impact on the world. Whereas the client-oriented cabinet member seeks to do his job to the satisfaction of the customers, the specialist cabinet member's internal definition of success depends on satisfying the norms of performance that the agency itself and its associated professions generate.

Since most federal bureaucratic agencies have histories, they also have built up over the years a population of leaders and former leaders, for example former assistant secretaries grown older, from which a president may draw members of his cabinet. A characteristic cost of this sort of leadership is a tendency for experts to know where too many of the agency's bodies are buried. Their knowledge gives them the means to use the agency's basic political capital if they succumb to the temptation to sacrifice the long-term interests of the agency in behalf of short-run goals or alliances that will make their incumbency of office look good. An advantage of a strategy of appointing specialists is that specialists are frequently highly competent managers, know the strengths and weaknesses of the organization intimately, and can draw the best work out of their former colleagues of the permanent government.

A third possibility is to call upon the Washington careerist. There is a very large group of possible cabinet officials who always live in Washington, whether or not they hold public office, and whose main expertise is general knowledge about how the federal government works within the Washington community. When these people are not serving in government, they may frequently be found in Washington law firms, public relations organizations, journalistic enterprises, or think-tanks. It is common for Washington careerists to associate themselves predominantly with the fortunes of one political party or another, but they only rarely have grass roots political knowledge or experience. Rather, these are typically bright graduates of the best universities who wash into government at an early age on the tide of some fresh incoming administration, and then stay around Washington once their tour of duty is ended, giving general advice—some of it very expensive—about how to work the American political system.

Frequently senior people in this group are considered eligible for a wide range of cabinet positions when a new cabinet is being formed, and in this sense they are interchangeable. They may have only tenuous ties to the electoral base of their political party and may or may not maintain special expertise with respect to the operations of one or another of the great bureaucracies. What they do offer is access to and knowledge about the byways of the Washington community and the modes in which it functions. Such officials offer general managerial skills, quite frequently of a high order, and the capacity to do policy analysis.

The great cost to a President who chooses a cabinet member from this population is the difficulty he may encounter in interpreting such a cabinet official's resistance—when it occurs— on substantive policy. When a client-oriented cabinet member disagrees with the White House, he can be presumed to be sending a message from a meaningful slice of the population at large, and it is a foolish President who ignores such a signal. Regardless of whether the President yields or persists in his line of policy, the signal means something to him, informs him about real political risks and costs. The specialist or expert is especially capable of vibrating to the pitch of his agency, and signs of disagreement from such a source may be the best early warning available to a President that a proposed plan of action will not work in a technical sense. A Washington careerist, however, may be vibrating only to the siren song of his own career. A Washington careerist's career may well entail longer time horizons than can be useful to the incumbent President, who must mobilize his influence and that of his administration so that it can be applied—spent, invested, even dissipated—before his term runs its course. Or the Washington careerist may simply be substituting the conclusions of his own policy analysis for an analysis in favor at the White House—or farther down in the bureaucracy.

On occasion, a Washington careerist can perform a significant service for a President by presenting to him the massed opinion of one or another policy Sanhedrin of resident senior Washingtonians that a contemplated plan of administration action will have highly damaging consequences for the professional reputation of a President, or for the welfare of the party or nation, or all of the above. This, evidently, was the great contribution of Washington careerist Clark Clifford, newly appointed Secretary of Defense, toward the termination of the Vietnam war. Clifford, with his delicate sense of touch in these matters, orchestrated a highly explicit show-and-tell session to persuade Lyndon Johnson that further prosecution of the war was futile.[19] The fact that this was needed for Johnson, a man himself so thoroughly embedded in the Washington milieu, is a very strong indication that resistance from a Washington careerist to a President's plans is normally inscrutable to the President.

A fourth option available to a cabinet-building President is to choose presidential ambassadors. These are officials who arrive in office principally because they are pre-presidential friends and allies of the President. Such officials frequently hail from the President's home state and—although there are exceptions—typically return there after their service with their particular President concludes.[20] These officials may be counted upon to respond with special alertness to presidential priorities, plans, and orders and to carry the President's message both to the agency and to its client groups. The rationale for a cabinet dominated by executives of this stripe is already familiar. Such executives are strongest in defending the presumption that the President, being the most recently elected chief executive, is entitled by his electoral mandate to command the resources of the executive branch and to shape its program according to his desires. The careers of presidential ambassadors are tied not to interest groups or to agencies or to the Washington community, but to the President personally.

Until the beginning of President Nixon's second term, commentators were not terribly alert to the possibility of pathology, cost, or difficulty in the pursuit of this alternative. Yet Presidents, it now appears, can ask members of the executive branch to do illegal or at least questionable acts, acts not contemplated and in some cases prohibited by the charters of the agencies involved. A President or his agent can seek to close down activities provided for by law or can repudiate political alliances with devastating future effects for himself, his party, and/or his successors. Those executives who are without expertise or independent standing with interest groups, the bureaucracies, or the Washington community are presumably least well situated to resist these tendencies when they appear.

To some observers it may seem odd and inappropriate that resistance to a President's expressed desires might occur as an administrative issue at all. This view, for example, seems to infuse the incredulity with which President Nixon's former speechwriter William Safire treats the difficulties Mr. Nixon had in getting World War I "temporary" government buildings cleared off the downtown Washington ceremonial parkland known as the Mall.[21] As Mr. Safire describes it, Mr. Nixon made an essentially aesthetic judgment. Flying overhead in his helicopter, it offended Mr. Nixon's sensibilities that the Mall had never been restored to its pristine state after the incursions of hasty wartime construction half a century before. So he ordered the buildings demolished forthwith. In fact, it took several years, a matter of great consternation to Mr. Nixon, and also, it appears, to Mr. Safire, who uses this as an example of the mindless bureaucratic inertia with which all Presidents must cope. Yet an off-with-their-heads proclamation is all very well when there is nothing more complicated involved than seeing that a glass of Fresca appears at the touch of a button on the President's desk.[22] A blazing fire can always be lit in the presidential fireplace if the President insists—as Nixon did—regardless of the temperature of the real world, since the presidential air conditioning can see to the overall comfort of the incumbent.[23] But Mr. Nixon made no constructive suggestion about what to do with the government employees who worked in the buildings he wanted obliterated, their files, equipment, office furniture, or functions. As luck would have it, given the vagaries of the Washington market in commercial real estate, it took time and not a little inconvenience and effort on the part of others, who also presumably had their ordinary jobs to do, to cleanse Mr. Nixon's vista.

It is, in short, one thing for a President to sit atop the administrative machinery of the government, blending and harmonizing the various contending interests that clamor for attention. The outcome is always bound to entail some administrative friction, some dissatisfaction, some reluctance, foot dragging, disagreement, and so on. In such circumstances there is no doubt that a President's political judgments ought to prevail. And sometimes, given the expenditure of sufficient resources, it does.[24] But when a President is merely taking his prerogatives, real or imagined, out for a morning stroll, it should come as no surprise when he has difficulty getting his way. Presidential ambassadors are likely to be the least well equipped of all his aides to see the distinction.

Finally, a President may choose his cabinet from representatives of symbolic constituencies. These are interest groups having little or no particular claims as organized clientele of the agencies to which their representative is appointed, but which do command the sympathetic attention of the press or constitute important communities at the electoral level. Symbolic constituencies are, in the United States, preeminently status groups and most commonly ethnic groups. Traditionally (but not entirely), ethnic groups have clustered together in the neighborhoods of America's eastern and midwestern cities, and in the party politics of these localities leaders of communal groups have demanded, and received, the recognition of nomination for or appointment to public office. This has been a common means by which parties have mobilized the loyalties, the precinct work and the votes of members of ethnic minorities, and has been identified as a device for accomplishing the assimilation of immigrant Americans into the mainstream of the American economy and society—without necessarily attenuating their communal identifications.[25]

The great advantage that representatives from symbolic constituencies bring to a cabinet is presumably the consolidation of the party loyalties of the communal groups from which they spring; the disadvantage is that they may have no great competence at or interest in doing the job at hand.[26] They may have no desire to learn the needs of their agency's clientele, or to understand the workways of the bureaucracy, or the impact of its programs, beyond the payoffs available to the cabinet official's own constituency.

3. Cabinet Building in the Carter Administration

No President is likely to pursue a pure strategy of cabinet building; rather, the strategy is typically mixed so that cabinet members may satisfy a variety of the criteria implied by the pure types mentioned above. Moreover, individual cabinet members may fit more than one category. Even so, the overall character of the mixture at any given time can be read as an expression of the claims to legitimacy made by each incumbent administration as well as of its dominant administrative style. This was certainly true of President Carter's administration, as a consideration of the seventeen cabinet-level appointments he initially made will illustrate.[27] In matters of foreign and defense policy, with the very important exception of United Nations Ambassador Andrew Young, Mr. Carter originally sought subject matter experts.[28] Cyrus Vance, Zbigniew Brzezinski, Michael Blumenthal, and Harold Brown had all put in time working in Washington for the government on problems associated with the departments they were asked to head.

In two instances Mr. Carter picked cabinet members who could be considered ambassadors from interest group constituencies. One, Bob Bergland, supervised the department whose interest group constituency was closest to Mr. Carter's own in private life, agriculture.[29] The second was the former governor of Idaho, Cecil Andrus, who brought to the Interior Department close ties with environmentalist groups.[30]

An academic labor economist, Juanita Kreps, served as the chief link between the business community and the administration: clearly not an interest group ambassador—at least not from the interest groups served by her department, but rather, widely proclaimed as a female appointee, hence a representative of a

Table 3.2 President Carter's cabinet, 1977

	Specialists	Client-oriented	Pres. Ambassad.	Washington careerist	Symbolic
State	Vance				
Treasury	Blumenthal				
Defense	Brown				
Justice			Bell		
Interior		Andrus			
Agriculture		Bergland			
Commerce					Kreps
Labor	Marshall				
HEW				Califano	
HUD				Harris	(Harris)
Transportation				Adams	
Energy				Schlesinger	
CIA			Turner		
Natl. Sec. Council	Brzezinski				
OMB			Lance		
Council Econ. Ad.	Schultze				
UN Ambassador			(Young)		Young

symbolic community.[31] Three lawyers, all Washington careerists in one way or another, who might well have been interchanged (as indeed one of them later was) headed the main
urban departments of Housing, Transportation, and Health,
Education and Welfare.[32] One of them, Patricia Roberts Harris
of HUD, as a black woman, held a status as a "twofer" among
symbolic interest groups that made her all but immune from
dismissal, as later events were to show.[33] There were, as many
people remarked, three Georgians, presumably presidential
ambassadors,[34] among the top seventeen as well as one former
classmate of the President's from the naval academy.[35]

Observers noted the odd resemblance between this Democratic cabinet and the Republican cabinet that immediately pre-
ceded it.[36] There was a curious neutrality of the Carter cabinet
toward the vast stew of interest groups, both within and outside
the government, that make up the traditional Democratic coalition. Of Mr. Carter's top seventeen appointees, how many
reached into the constituencies suggested by the old New Deal

voting coalition, indeed the coalition which came together to elect him?[37] Where were the representatives of the Irish, the Polish, the Jews, the Italians, the cities, the labor unions? Where indeed were the long-time active members at the grass roots of the Democratic party? Not wholly absent to be sure, but hard to find. The Carter cabinet was far stronger in symbolic than in traditional interest group representation. National Security Council Chief Brzezinski was in fact of Polish extraction, and even spoke English with a slight Polish accent. Yet it could be argued only in jest that this Columbia University professor "represented" Polish-Americans in the cabinet, unconnected as he was with the great Democratic political organizations in the eastern and midwestern cities in which Polish-Americans predominated. The same was true for Secretary of H.E.W. Joseph Califano and Italian-Americans, and so on, with only a couple of exceptions, the most significant once again undoubtedly being Andrew Young and the black community.

So, perhaps the first clear signal given by President Carter in the way he constituted his administration was to proclaim a disbelief in the reality of the interest group composition of the Democratic party. He understood the need for symbolic gestures, and for satisfying those interest group demands made through the mass media, but there was, clearly, nothing in his experience of the national Democratic party—principally, by then, revealed to him by the nominating process—that confronted him with most of the varied components of the grand Democratic coalition. His own faction—the enthusiasm of his "peanut brigade" of campaign workers, and of Carter voters in primary elections—is what nominated him. Once nominated (at a national party convention paid for in full by the government) the federal government subsidized his entire general election campaign. Although it was true that his fellow-Democrats elected him President, this fact was easy to ignore, and Mr. Carter remarked on entering the White House that he "owed" his presence there to nobody.[38]

His cabinet appointments, strongly weighted with can-do

technocrats, leavened by gestures to symbolic constituencies, and filled out with Georgians, proclaimed some of the implications of the President's viewpoint. Other consequences were to emerge in President Carter's dealings with Congress.

4. Getting Along with Congress

A second significant piece of evidence showing how the lessons of the process that nominated President Carter were permitted to override imperatives arising out of the post-inaugural Washington environment emerges from a consideration of President Carter's relations with Congress. These were very unpleasant.[39]

Outside observers were entitled to view Mr. Carter's inability to get along with Congress with some amazement. After all, President Carter was a middle-of-the-road Democrat, and the Congress was controlled—overwhelmingly controlled—by middle-of-the-road veterans of the Democratic party. After the election of 1976, some 292 Democrats were returned to the House of Representatives (out of 435) and 62 Democrats out of 100 sat in the Senate. The midterm election of 1978 gave the Democrats a 276 to 159 advantage in the House and 59 to 41 in the Senate. The last time an overwhelmingly Democratic Congress had coincided with a Democratic President, after the 1964 landslide election, the result had been the bumper crop of new and innovative programs of the famous 89th Congress.[40]

Times had changed on Capitol Hill since the enactment of the Great Society, however. Friends of President Carter were quick to point out that important alterations had overtaken Congress in the intervening decade and a half, making the Capitol Hill tasks of a President—any President—far more difficult. Congress had democratized its rules, for example, had endured a period of fierce antagonism with President Nixon, had created an enormous staff bureaucracy in part to wage war on the executive branch, and had undergone a drastic turnover in membership so that a majority of members could not

hark back to the good old days of presidential-congressional
cooperation.[41]

In the old days, it was said (the examples frequently coming
from the Kennedy era of 1961–63), a President could strike a
bargain with the congressional leadership, and the leadership
could deliver the Congress. Committee chairmen ruled their
roosts, and this meant that once a committee chairman committed himself to cooperation with the President, the President's task of assembling a majority in Congress was greatly
simplified. As Lance Morrow said in *Time:*

> Congress used to operate through party discipline enforced by
> powerful leaders like Sam Rayburn, who in turn responded to
> leads from the White House. Now Congress has become a catfight
> of centrifugal energies, a fractured, independent crew that in its
> less disciplined moments approaches the opera buffa standards of
> the Italian Chamber of Deputies. As never before, Congressmen
> have narrowed their definitions of their responsibilities: they an-
> swer to their constituencies and to their special interests. Those
> arrayed demands do not necessarily correspond to the national
> good.[42]

In Mr. Carter's time, the argument continued, committee
chairmen had lost their power to the chairmen of subcommittees, and subcommittee chairmanships were dispersed among
the multitudes. Instead of cultivating an alliance with twenty
senior Congressmen, presidential legislative liaison workers had
to court 120 subcommittee chairmen, a much more complicated task.

There is a grain of truth in this argument, but no more than
a grain. The "good old days" existed only during the brief span
of the 89th Congress, well after Sam Rayburn had gone to his
reward.[43] And the bad new days of the Carter era were structurally far more favorable to a middle-of-the-road Democratic
President than anything Presidents Truman or Kennedy ever
saw. Congressional reform devolved power not only downward
to subcommittee chairmen but also upward to the House Democratic leadership. Mistakes, ineptitude, and presidential ne-

glect of Congress played a far more significant role in creating the Carter administration legislative record than Carter administration apologists admit.

The litany of presidential mistakes toward Congress was nearly endless: an inability to settle upon legislative priorities, a reluctance to bring Congress into the process of formulating proposals before they arrived, fully blown, on Capitol Hill, a disinclination to interact or bargain directly with Congressmen and a tendency to appeal to a mythical entity known as "the people," presumably over the heads of Congressmen, themselves elected public officials, the vast majority of whom had run well ahead of Jimmy Carter in their home districts.[44]

As two junior sub-cabinet Carter appointees recalled:

> The President sent a flotilla of major proposals to the Congress in the first eighteen months of his administration—cuts in water projects, social security finance, a comprehensive energy program, a tax rebate scheme, hospital cost containment legislation, comprehensive tax reform, welfare reform. Many of these proposals went to the tax-writing committees of the Congress: Senate Finance and House Ways and Means. And because the President had overloaded the Congress and those committees with reforms that would not command ready assent, because he was not able to marshal the administration's resources and develop political support for all the major battles that were required, and because many of his top political lieutenants were untutored in the ways of the Congress (and at the outset didn't seem to care), most of these proposals were either sunk or badly damaged. President Carter's reputation as less than skillful in domestic affairs and with the Congress was thus firmly established.[45]

President Carter's legislative liaison was in the hands of a person totally inexperienced and unknown on Capitol Hill, but this was only the beginning of the problem.[46] The Carter administration could—and eventually did—hire people more experienced in the ways of Congress to join what in time became a bloated liaison staff. For a while many of these people were housed in the White House's East Wing, physically and symbolically removed from the West Wing center of presidential power, and only Frank Moore, their chief, ever gained much direct access to the President.[47] The Carter administration ignored the advice of friendly predecessors to establish regular beats for liaison personnel based on the bloc structure of Congress and instead began by assigning issue specialties to them.[48] This meant that different liaison people would deal with the same Congressman on different issues, and no regular relationship, no orchestration of give and take over the long haul, could easily be established. Worse, presidential liaison people, allegedly issue specialists, were never tied into the policy formation process in the White House and so were denied both flexibility and credibility in dealing with Congress.

Despite repeated efforts from congressional leaders to bridge the gap between Capitol and White House, neither President Carter nor his closest aides who actually participated in policymaking made informal acquaintances in Congress. It became common coin on the Hill that Mr. Carter had conceptualized Congress as indistinguishable from the Georgia legislature.[49] Democrats from all parts of the political spectrum—but most notably those from his own part—began to collect and disburse, like children with bubble-gum cards, a fund of Jimmy Carter stories illustrating his utter lack of interest in listening to congressional advice, his stubbornness, his parochial insularity. He evidently had no back-channels to Capitol Hill and wanted none, no congressional cronies, no unofficial sources of information, indeed virtually no friends. A *New York Times* news analysis examined the phenomenon after more than two years of the Carter Presidency:

> The result, simply, is that he has no friends on Capitol Hill. "He just doesn't have that wellspring to call on," said one Southern Senator. "When you get in trouble, that's when you need your friends to come to your defense, and he just doesn't have that."[50]

For a President to "get along" with Congress and to "have friends" there is not merely a matter of bonhomie or backslapping, though superficial tokens of affability are generally appreciated in the Washington world as they would be in any situation. Where there is considerable and constant turnover in the cast of characters that politicians must do business with, a pleasant disposition and a willingness to learn to do business on short acquaintance is doubly advantageous. More fundamentally, however, getting along with Congress means such things as showing respect for the institutional legitimacy of what is after all a constitutionally sanctioned branch of government, learning enough about the politics of the place to be able to draw with discernment on the talents and abilities, wisdom and good will of potential allies in shaping a reasonable policy agenda, and understanding in a non-censorious way something about the political problems and ambitions of members. Leadership by a President of an institution like Congress is certainly not precisely like the leadership of a strict hierarchy by the person at the top of the organization chart. But both sorts of leadership—especially congressional leadership—involve instilling an atmosphere of cooperation, of recognizing that credit must be spread if responsibility is to be spread, and that the morale of the troops may make a difference to overall productivity.

President Carter's difficulties on all these counts surfaced very early in his administration. A month after inauguration day the Speaker of the House told Elizabeth Drew:

> The problem with the people around Carter is that they spent so much of their time running against Washington they don't know they are now a part of Washington.[51]

And another of her sources said:

> He spent so much time in the campaign saying that he didn't owe anybody anything that nobody thinks they owe him anything.[52]

Mr. Carter's press secretary, Jody Powell, said after six months in office:

> It's the damndest thing about him . . . He went all over the country for two years asking everybody he saw to vote for him for President, but he doesn't like to call up a Congressman and ask for his support on a bill.[53]

The political resources of such a President were bound to be easily depleted, even among those, undoubtedly a great number, who wished him no particular ill. Given heavy Democratic majorities, the Speaker and the Senate Majority Leader could rally majorities in behalf of a President who could not or would not help himself, but not easily and not often. For since the rise of large congressional staffs, Congressmen no longer need to take the word of the executive branch on any controverted point if they choose not to. The congressional party can now, if it chooses to do so, chart its own course with respect to policy fully in possession of adequate intellectual fortification. The capacity to do so, however, operates at least in part independently of the inclination to do so. The development of staff capacity was a congressional response to hostilities during the presidency of Richard Nixon.⁵⁴ The use of this capacity when Congress and the Presidency were both securely in Democratic hands was a response to the presidential style of Jimmy Carter.

In order to get a sense of the internal dynamics of the Carter Congresses in relation to the Presidency some historical background is helpful. Essentially, throughout most of the last fifty years, the internal struggle of central importance to Congress has been between liberal and mainstream Democrats on one hand and a conservative coalition, encompassing conservative Southern Democrats and mainstream Republicans, on the other.⁵⁵ Franklin Roosevelt's abortive attempt in the election of 1938 to purge Congress of conservative Democrats after the failure of his Court-packing plan was an early recognition of the capacity of the conservative coalition to hamper the political plans of liberal Democratic Presidents.⁵⁶ The strength of these two grand coalitions has ebbed and flowed over the years, changing tidally with the results of biennial elections. Both sides have taken what advantages they could from internal rules of Congress pertaining especially in the Senate to freedom of debate, in the House to control over the agenda, and in both houses to seniority in committee assignments. And both have interacted strategically with the President—the conservative group mobilizing their strength, when they needed to, mostly around threats of vetoes by conservative Presidents, and the liberal group around the priorities of liberal Presidents' programs. The great resources in the hands of the conservative coalition over most of the fifty years have been the seniority of their leaders, and their tactical skill and tenacity. The great resources of the liberal group were numbers, and the claims on the Democratic side of party loyalty.

By 1960, the liberals had the numbers in the Senate, but in the House of Representatives the picture was quite different. The election of 1958 was a Democratic landslide year, sending 283 Democrats to the House.[57] Yet, as Representative Clem Miller, a liberal Democrat from northern California, pointed out at the time in one of his brilliant newsletters:

> The combination of Southern Democrats and Northern Republicans can always squeak out a majority when they want to, and they want to on a great number of significant issues . . . Actually, the Democratic party as non-Southerners define it is a minority in the House.

"There are 160 Northern Democrats and roughly 99 Southern Democrats," Miller figured:

> This includes Texas, but does not include the border states . . . which generally cancel each other out. Maryland votes against us, West Virginia with us, Missouri cancels itself out, half liberal, half Southern, and Kentucky, ambivalent, sometimes with us and sometimes against us . . . Begin with a base of 160 Northern Democratic votes. Add to it fifty percent of the (roughly 30) border state Democrats. We are always 15 to 30 votes shy. . . .[58]

So long as the conservative Democrats from the South had the option of coalescing on the floor with a mostly united Republican party, efforts to organize the Democratic party in the House by requiring greater party loyalty of such a large minority were bound to come to grief. From 1958 to 1978, however, the strength of this minority within the Democratic party in the House eroded so as to permit effective action in the Democratic caucus by the end of that period. Meanwhile, liberal Democrats became better organized through the formation of the Democratic Study Group and were better able to mobilize their big battalions as needed.[59]

A slightly different way of doing Clem Miller's arithmetic yields the twenty-year comparisons visible in Table 3.4.

Over a twenty-year period it was mostly conservative southern Democrats who lost their seats to Republicans.[60] Outside the South, Democrats, mostly liberal, replaced Republicans. So over the twenty-year span the House became more liberal overall, and this trend was accelerated within the Democratic caucus.

Table 3.4   Coalitions in the House, 1960 and 1980

	86th Congress 1959–60 Elected 1958	96th Congress 1979–80 Elected 1978
Democrats	280[a]	281
Republicans	152	154
Southern[b] seats	106	108
Conservative Democrats	66	47
Mainstream[c] Democrats	33[d]	31
Republicans	7	30
Non Southern Seats		
Democrats	181	203
Republicans	145	124
Democratic Caucus		
Non-South plus		
Mainstream South	214	234
Conservative South	66	47
Conservative Coalition		
Republicans plus Conservative		
Southern Democrats	218	201

[a] Three vacancies by the end of the Congress
[b] Southern seats are seats from 11 states of the old Confederacy.
[c] Mainstream Southern Democrats are those whose CQ party support scores exceed their party opposition scores by at least two to one.
[d] Includes Speaker Rayburn

Sources: *1960 Congressional Quarterly Almanac* (Washington, D.C.: Congressional Quarterly, 1960), pp. 140–41; *Congressional Quarterly Weekly Report* (January 10, 1981), pp. 82–83.

Indeed, the Democratic caucus was the engine of change within the House during the decade 1970–80. Prodded by its organized liberals, the caucus established a subcommittee bill of rights, took power from committee chairmen, deposed chairmen in historic breaches of seniority, and put the Speaker in charge of committee assignments in general and of assignments to the Rules Committee in particular.[61]

Some of these changes clearly decentralized power; but some took powers previously dispersed to committees and their chairmen and vested them in the House Democratic leadership, and especially the Speaker. And it was an organ of centralized party leadership, the Democratic caucus, that did it.

These observations must be kept in mind in evaluating the claim that Congress became less tractable than previously to leadership from a Democratic President. The twenty-year record of the institution suggested, rather, sizable gains in the numbers of regular Democratic members, and an increased potential for favorable results for any Democratic President willing to work with the congressional leadership in establishing legislative priorities and strategies. A proliferation of subcommittees meant, after all, that guidance was needed in scheduling the orderly floor consideration of what would otherwise soon become an indigestible log-jam of proposals. And with his exclusive right to make appointments to the Rules Committee, the Speaker gained the influence he needed to coordinate traffic headed for the House floor. This influence was denied all Speaker O'Neill's predecessors, reaching back to the revolt against Joseph Cannon at the turn of the century.[62]

So it will not do to argue that the undeniably significant changes in the way Congress did business were at the root of difficulties that President Carter had in mobilizing congressional support for his proposals. Congressional change was not a cause of President Carter's problems with Congress, and more generally in governing. Quite to the contrary, if anything, a hard look at Congress deepens the puzzle of Carter's difficulties. By any reasonable gauge, relations with Congress ought to have been a bright, not a dark spot in President Carter's record. And so, far from being an explanation, President Carter's difficulties with Congress themselves need explaining.

An appropriate explanation is readily to hand, for, among the side effects of changes in the presidential nomination process, as we have seen, have been sharp reductions in the number of Congressmen who participate. The numbers for the Democratic party are startling:

Table 3.5 Participation in Democratic National Conventions

	Percentage of Democratic U.S. Senators who were voting delegates or alternates	*Percentage of Democratic U.S. Representatives who were voting delegates or alternates*	*Percentage of Democratic Governors who were voting delegates or alternates*
1956	90%	33%	100%
1960	68	45	85
1964	72	46	61
1968	68	39	83
1972	36	15	80
1976	18	15	47
1980	14	15	76

For comparison, Republican National Conventions:

	Senators	*U.S. Representatives*	*Governors*
1968	55%	31%	88%
1980	63	40	74

Data were drawn from official Convention Calls for each year.

Sources: Commission on Presidential Nomination and Party Structure (Morley A. Winograd, chairman), *Openness, Participation and Party Building: Reforms for a Stronger Democratic Party* (Washington, D.C.: Democratic National Committee, January 1978); and Nelson W. Polsby, "The Democratic Nomination," in Austin Ranney (ed.), *The American Elections of 1980* (Washington, D.C.: American Enterprise Institute, 1981), p. 57. The Republican data were taken from James Ceaser, *Reforming the Reforms* (Cambridge, Mass.: Ballinger, 1982), chapter 3, footnote 18.

Any President who reads the Constitution can see how desirable it is to make common cause with allies in Congress.[63] It must be accounted a significant, indeed a massive incapacitation of party nomination processes that these processes should have operated to persuade a President otherwise to the extent they evidently did in President Carter's case.

5. *The Crisis of Confidence, 1979*

Finally, there are the lessons President Carter himself drew from the accumulating evidence of stalemate in government and his unpopularity among citizens at large as his term wore on. These too testify to the extent to which nomination politics was a model for the politics of governing in President Carter's thinking.

By the middle of the third year of his Presidency, Jimmy Carter's ratings in most public opinion polls were roughly comparable with the worst days of Harry Truman or with Richard Nixon on the eve of impeachment. In early July 1979 the *New York Times* reported that only 26 percent of Americans approved of Mr. Carter's handling of the Presidency, a slide of four points from the previous month.[64] To be sure nobody was talking about impeaching Mr. Carter, not least because he had committed no impeachable offenses. But Democrats from many points in the political spectrum were actively mounting a dump Carter movement designed to prevent his renomination, and substitute the far more popular Senator Edward Kennedy.[65] Indeed, the same *New York Times* poll showed 53 percent of Democratic respondents favoring Kennedy as the Democratic nominee as compared with only 16 percent who preferred Carter. And when Kennedy supporters were asked their next choice, more preferred Governor Jerry Brown of California than the President.[66] Republicans could take heart in the public opinion survey trial heats that showed two of their possible presidential hopefuls, Ronald Reagan and Gerald Ford, defeating Carter in an election.[67]

The President had to absorb criticism on many fronts: there was considerable inflation in the domestic economy, which economists believed could not be stemmed short of a recession; after several years of weakness against foreign currencies, the dollar continued unstable; the OPEC countries were raising the price of oil in large jumps and limiting production; a friendly regime in Iran which had supported many aspects of American diplomacy collapsed into hands that were intermittently unsteady and hostile; the price of gasoline crept toward a dollar a gallon and beyond, and sporadic shortages caused unexpected closings of gas stations, panic buying, and long lines, first in California, then further east.

David S. Broder of the *Washington Post* described Mr. Carter's response to these and to other signs of mounting difficulty for his Presidency in the early summer of 1979:

> Alone [at Camp David] with his wife, Rosalynn, Jimmy Carter sat down on the afternoon of July 4 to put his pencil to a draft of an energy speech scheduled for national television the next night . . . But between noon and 3:00 PM that Independence Day, away from his staff, Cabinet and administration, Carter decided to cancel the speech for one simple reason. He believed that neither the country nor the Congress would heed or respond to another energy speech—the fifth of his term—from him.[68]

For twelve days Mr. Carter stayed in his mountain-top retreat, leaving only twice by helicopter to sample opinions on the state of the nation from groups of private citizens: once on the back porch of a machinist's house in Carnegie, Pennsylvania, and once in the living room of a retired cattle farmer in Martinsburg, West Virginia.[69] During his twelve-day retreat, he summoned a varied cast of experts, religious leaders, labor union chiefs, selected officeholders and journalists—well over a hundred in all—to counsel with him.

Martin Schram of the *Post* reported:

> The thing that Carter and his top advisers have come to see is that the President, not just the nation, had lost the way.

> Carter found it on the mountain-top. He found it not just from listening to invited guests telling him what was wrong, but from long walks in the woods and from reading such writers as John Gardner on morale and James MacGregor Burns on leadership.

> "The President began to understand that the crisis at hand is not limited to a political crisis but [is] a leadership crisis," said one of that inner circle of advisers with him throughout his Camp David conference.[70]

Finally emerging from Camp David, he gave a nationally televised address partly on the nation's energy problem and partly
about what he described as a crisis of confidence sweeping the
United States.

> It is a crisis [the President said] that strikes at the very heart and
> soul and spirit of our national will . . .

> The erosion of our confidence in the future is threatening to de-
> stroy the social and political fabric of America.

Mr. Carter listed as symptoms of this crisis the following:

> For the first time in the history of our country the majority of our
> people believe that the next five years will be worse than the past
> five years.

> Two-thirds of our people do not even vote.

> The productivity of American workers is actually dropping . . .

> The willingness of Americans to save for the future has fallen
> below that of all other people in the western world . . .

> There is a growing disrespect for government and for churches
> and for schools, the news media, and other institutions.

Among the causes of the crisis, Mr. Carter identified the as-
sassinations of John and Robert Kennedy and Martin Luther
King, Jr., the lack of success of our armies in Vietnam, the
"shock" of Watergate, ten years of inflation, and "a growing
dependence on foreign oil." He then sounded a theme familiar
to those who recalled his successful campaign for the Presidency, just three years before:

> Looking for a way out of this crisis, our people have turned to
> the Federal Government and found it isolated from the main-
> stream of the nation's life. Washington D. C. has become an island
> . . . a system of government that seems incapable of action.

> You see a Congress twisted and pulled in every direction by
> hundreds of well-financed and powerful special interests.

> You see every extreme position defended to the last vote . . . by
> one unyielding group or another . . .

Often you see paralysis and stagnation and drift. You don't like it
and neither do I.[71]

The next day Mr. Carter gave two further speeches, in Kan-
sas City and Detroit, both of which were well received in the
mass media, as his public opinion rating bounded upward.[72]

A Carter aide told Martin Schram: "I think we have seen
both the rebirth of the American spirit that he talks about and
the rebirth of the Carter presidency as well." And Schram
added, "In style, scope, and effort expended, that is certainly
true."[73]

There then ensued a sequence of events that astounded most
political observers. At a special White House meeting of the
cabinet on July 17, Mr. Carter gave an hour-long lecture to the
most senior members of his administration on the failings of
his Presidency, which ended with his request for the resignations of all those present.

Partial reconstructions of this meeting leaked out for the next
month.[74] A cabinet officer, recalling the unhappiness when
Richard Nixon undertook a wholesale house-cleaning of the
White House staff at the beginning of his second term, asked
whether formal resignations were necessary from officials who
serve at the pleasure of the President anyway. There was debate over whether the offers had to be made in writing or could
be made orally. At some point the President excused himself
from the meeting, leaving his trusted associate, Hamilton Jordan, the only White House staff member present, in charge.
Jordan "stood up and slapped down a large stack of 'staff evaluation forms' on the cabinet table."[75] These thirty-item forms
were to be returned in 72 hours, and were to be used in pre-
paring evaluations of subcabinet officials whose jobs it was
thereby understood were also in jeopardy.

Questions on this form, meant to be filled in by cabinet officials about their senior associates,[76] included such items as: "On
the average when does this person arrive at work? leave work?"
"How bright is this person? (on a scale of one to six)," and "To

what extent is this person focused on "accomplishing the administration's goals? Personal goals?" These two alternatives, as the form indicated, were supposed to sum to 100%.

Jordan offered to go to the President for purposes of clarification with three options for those present: no resignations, oral resignations, or written resignations. The President sent word later that day that oral resignations would do. In a conference telephone call, members of the cabinet were told by Jordan that it would be publicly announced that the offers had been made.

Later in the day it became known that Jordan had also run a meeting of senior White House officials, who were asked to follow suit. It was announced that Mr. Jordan, who already possessed the corner office in the White House's west wing that went with the job, would assume the title of White House Chief of Staff.[77]

Two cabinet officials seemed to have been the intended targets of all this activity: Secretary of Health, Education and Welfare Joseph Califano, and Secretary of the Treasury Michael Blumenthal. Three other resignations were also accepted: Attorney General Griffin Bell, by amicable prior agreement, and Energy Secretary James Schlesinger, who had previously advised the President that he thought his political usefulness was nearing its end. Transportation Secretary Brock Adams became an additional casualty when he found himself publicly at odds with White House announcements during the process itself.

No one questioned President Carter's constitutional authority to make these changes, nor were any of them made—as had been the case with President Nixon's October 1973 firing of Archibald Cox—in violation of some informal political commitment or understanding with Congress. Nevertheless the immediate reaction on Capitol Hill from Democrats and Republicans alike was overwhelmingly unfavorable.[78]

The liberal Democratic Study Group in the House of Representatives put out a parody of Hamilton Jordan's report card, in which the White House staff (instead of sub-cabinet officials) was the subject. One high-level official said that Jordan's forms "were more appropriate for a junior high school student than an executive running a $40 billion program."[79] Another said: "This will make us the laughing stock of the world. Everybody knows who is performing well in this Administration . . . If those guys in the White House don't know it, they should be fired."[80] "All over the House floor, they're shaking their heads over that one," said moderate Republican Congressman Barber Conable. "It's very destructive to morale . . . the government has damn near ground to a halt."[81] Nearly a month later a high level administration official described congressional reaction to Jack Nelson of the *Los Angeles Times* as still "savagely angry. We are in deep political trouble."[82]

Unfavorable comment was not restricted to Congress. Virtually every articulate group of Washington political observers was tapped by the newspapers for reactions, and responses ranged from puzzled to contemptuous, with only a few voices raised in the President's defense. For example, Carter supporter Irving Shapiro, chairman of the DuPont Company, said: "I'm simply baffled by this procedure. It casts a cloud over the whole administration."[83] A high-ranking previously loyal member of Carter's own administration said: "At this point, I don't see how I could campaign for his reelection."[84]

Major newspapers printed reams of adverse comment in their news columns and editorials and editorial cartoonists had a field day. More seriously, foreign press reaction was also unfavorable, so much so that presidential press secretary Jody Powell and national security affairs adviser Zbigniew Brzezinski were moved to call in Washington-based foreign journalists in an attempt—mostly unsuccessful—to persuade them that nothing out of the ordinary had occurred.[85]

Foreign money markets made their own evaluation of the situation. There was a noticeable flight from the dollar abroad, and the price of gold floated over $300 an ounce for the first time in history. In two weeks, the dollar lost 70 percent of the ground it had gained on the West German mark since the dollar-rescue program was put into effect by the U.S. Treasury and the German Bundesbank in November 1978.[86]

The most puzzled reaction came from those who could not understand why it was necessary or prudent to ask for thirty-four resignations in order to accept a handful of them. What was characterized by the President and by his aides Hamilton Jordan and Jody Powell on national television as an "orderly, methodical" process could have easily been foreseen to produce feelings of unease in foreign capitals as well as in Washington and at least momentarily to jeopardize such important long-term administration goals as stabilization of the dollar, ratification of the SALT Treaty, and the administration's energy proposals in Congress.

Adverse comment was thus compounded of a number of elements. There were those, chiefly Republicans, who were ready to take a simple pleasure in the obvious disarray of a Democratic administration.[87] Others, chiefly Democrats, had come to admire one or more of the departing cabinet members, or disliked or distrusted many leading—and surviving—members of Mr. Carter's White House staff.[88] Equally important were the critics who fell into neither category, but consisted of those for whom the manner in which the shake-up was conducted showed an appalling lack of respect for the top management of an administration that was mostly blameless by White House standards, and a failure to anticipate the disruptive consequences caused by the wider ripples of concern that were sure to flow from the staging of such an unnecessary and unnecessarily dramatic event. One observer said: "One does not stage the entire fifth act of *Hamlet* to get rid of Laertes alone."[89]

The most sympathetic rendering of President Carter's thinking ran along the following lines: He had become genuinely distressed by the data and the analysis contained in what was described as a "voluminously dire memo" presented to him in April 1979 by his public opinion analyst Patrick Caddell.[90] As his speech indicated, these findings showed Americans to be pessimistic about the future, for themselves and for the nation. Mr. Carter felt a presidential obligation to seize the moment to lead, and felt he must put his own house in order to do so. Thus the retreat to Camp David, the wide consultations, the wholesale request for resignations to symbolize a new start. Members of the administration who had not proven themselves to be team players could be removed without excessive personal recrimination, and other, borderline cases could be put on notice that a new degree of commitment would be required of them.

It must be said that virtually nobody seems to have found this set of explanations believable. Too much other data crowded in and demanded to be taken into account, such as Mr. Carter's own low standing in the polls, the shaky empirical foundations of the Caddell analysis, the gap of three months between the delivery of Caddell's memo and Mr. Carter's decision to act on it.[91]

There was also a highly projective flavor to Mr. Carter's response, which entailed a pledge to reorganize his time so as to perform fewer merely managerial functions and to travel more out among ordinary Americans. This plan of action entirely neglected the heavily populated circle of intermediaries who are themselves organized for the purposes of sharing in the governing process and representing varied segments of the population in their relations with the government. Mr. Carter's view of "special interests" and of Congress was clearly set forth—and not for the first time—in his July 15 speech. They, the press corps and the great bureaucracies of the executive branch all together made up the "Washington community," who, Mr. Carter insisted, were out of touch with popular sentiment and hence could be ignored. The Washington press corps, in Mr. Carter's opinion, took a much dimmer view of his cabinet shake-up than the people at large, which both *Newsweek* and the *New York Times* reported, in thinly disguised indirect discourse, "fortifies his decision to hold fewer of his regular Washington press conferences and get out around the country more in the meet-the-people style of his 1976 campaign."[92] As the *Times* reported:

> Jimmy Carter reached a number of conclusions about the way he was handling his Presidency. One was that he was wasting his time, or about 40 percent of it.
>
> That is the percentage of . . . time that was committed . . . to regularly scheduled weekly meetings with budget and staff aides, department heads and the like . . .
>
> Under the new arrangement, Mr. Carter is to be free to leave the Oval Office more frequently and travel outside Washington at least once a week. He has told visitors to the White House that he intends to move around the country as much as possible in an effort to reestablish the direct contact with the public he had during the 1976 campaign.[93]

On "Meet the Press" (NBC) Hamilton Jordan was asked why the administration had proceeded in the way it had and replied that "The American people are not concerned as to exactly how it happened, but . . . with what has happened."[94]

Observers began to ask if all the thrashing about was an effort—admirable or otherwise—by the President to position himself so as to be able to run against Washington for reelection to Washington's most prominent job just as he had originally run as an outsider in 1976. Could all this turmoil merely be an attempt to create an administrative tool in the spirit of zero-base budgeting, the moral equivalent of war, and other examples of tabula rasa thinking much favored by the President[95] that would aid him in shoring up perceived political or managerial weaknesses in his administration? Questions of motive and intention were regularly confounded with questions about consequences and all sorts of answers were readily to hand. Because these answers were not always free of partisan interest and because many key items of information had to be inferred or guessed at, a fair amount of confusion remained.

Mr. Carter's own diagnosis, however, seemed clear enough. He pledged to reform himself and his Presidency by re-creating

the conditions under which he was originally nominated: going
out more into the country, making contact with citizens and
leaving the day-to-day management of the government in the
hands of Hamilton Jordan and his newly reconstructed cabinet,
which to a far greater degree, he believed, would reflect the
priorities and the policies of the Carter White House. This
seemed to be an unequivocal statement of a belief that the central failure of the Carter administration was one of public relations, not, as so many Washington observers insisted, one of
governing. This latter diagnosis of his position emphasized that
in the thirty months of the Carter Presidency up to then, Mr.
Carter had not learned how to get along with Congress, nor
with the congeries of interest groups clustering in the nation's
capital and organized to do business there.

According to this view, while Mr. Carter had correctly noted
his low ratings in the public opinion surveys, their cause was
not a lack of preaching and teaching from the President to the
nation at large but rather reflected an accumulated sense of
malaise and difficulty which were the product of Mr. Carter's
own disinclination to cooperate with other political leaders in
Washington. These leaders—leaders, for example, of the labor
movement, leaders of Congress both in the Democratic and the
Republican parties—in turn fed back to their own constituencies throughout the country their dissatisfactions, and the cu-
mulative effect of these dissatisfactions had finally adversely affected Mr. Carter's ratings in the public opinion polls. This close
paraphrase of the theory of the two-step flow of public opinion
formation suggests that even though it is now possible for political leaders to reach out and influence the opinions of mass
publics directly through the medium of television, over the long
run people also look to leaders more proximate to themselves
than the President for a sense of political orientation.[96] Thus,
according to this argument, Mr. Carter's inability to win over
and to learn to cooperate productively with members of Congress and leaders of interest groups—especially those near to
him on the ideological spectrum—had taken their toll in the
public opinion polls by an indirect route.

If this diagnosis had any merit at all then it meant that Mr. Carter had already, as he originally suspected on July 4, come perilously close to saturating his capacity to influence public opinion in his own favor directly. Thus the best method available to him to turn around his low ratings in the public opinion polls was not to travel about the United States, but to face trying to do something about his problem of getting on with the Washington community. His analysis of the situation and his pledge to remedy it pointed in exactly the opposite direction, however, and suggested that he had overlearned the lessons of the nomination process. These foretold that he would continue his nearly bankrupt strategy of neglecting if not attacking Congress and the interest groups, even those clustering around the Democratic party, and attempting to reach out directly to the people over the heads of these other political leaders in order to become a more popular President.

Even if he could become more popular by this route, it was uncertain how it would help him get greater cooperation from Congress. It was also uncertain how it would improve the prospects of the various legislative programs that he had in mind. At best a high standing in public opinion is a resource that a President may call upon in order to persuade legislators of the popularity of the overall program that he advocates. No sophisticated legislator is likely to suppose that there is a genuine one-to-one correspondence between a President's popularity and the popularity of any particular item on his agenda, but members of Congress are far more inclined to give a popular President the benefit of the doubt in considering a presidential proposal than they will a President of marginal popularity or of low popularity.

In any event a President's popularity is only one of a number of considerations which legislators generally feel that they must take into account in determining their own opinions about public policy proposals.[97] They must also concern themselves with their popularity with their own constituencies. They may have their own ideas on the merits of issues as a result of personal knowledge or experience or their access to congressional sources of information or the play of national interest groups upon them both in their constituencies and in Washington itself. Presidents can participate in orchestrating the deployment of forces upon individual members of Congress, upon their own congressional party and upon the Congress as a whole, and therefore cultivating popularity with public opinion at large is one of a number of resources and instruments available to a President who means to conduct a successful legislative program.

Yet Mr. Carter initially gave no indication of an intention to get on more happily with the Washington community. He went out of his way to speak ill of special interest groups as a general category and made no concession to the idea that some small part of the national interest might be found lurking in a corridor on Capitol Hill. Even more significantly he made no attempt to reorganize his congressional liaison operation, and although many special assistants to the President in the White House were required to proffer their resignations, not one was immediately accepted. The only material change that took place was Mr. Jordan's de jure elevation to the position of Chief of Staff.[98]

Mr. Jordan, undeniably a person in whom the President reposed great confidence, was nevertheless in very bad odor on Capitol Hill. It was said by Congressmen that Mr. Jordan was harder to reach on the telephone than the President himself.[99] The Senate Majority Leader—a Democrat in good standing—professed not to know him at all.[100] The Speaker of the House said that in the thirty months of the Carter administration he had only met Mr. Jordan twice.[101] Many other comments unfavorable to Mr. Jordan—not all of them on only slight acquaintance—also greeted the news of his elevation.

Outgoing Attorney General Griffin Bell, for example, a genuine card-carrying member of the Georgia Mafia, was reported as "telling friends . . . in a round of farewell parties that Hamilton Jordan . . . 'is not long on maturity' and might be taking on too big a job."[102]

"I am not well known and I recognize that," Mr. Jordan said

at what he described as his "debutante party," a small cocktail buffet convened on July 27, 1979—roughly two and one-half years after Jimmy Carter's inauguration—to give the Speaker and twenty members of Congress a chance to meet him. "Most Congressmen that are criticizing my elevation to Chief of Staff don't know me." Mr. Jordan continued, "That's not their fault; that's my fault. I have not made the effort to know them."[103]

All this activity constituted an open acknowledgment that the Carter Presidency was experiencing great difficulty. A question remained whether the difficulty was merely a matter of tell-tale inefficiencies, or a not-quite-tight-enough ship, or sartorial sloppiness among the White House palace guard,[104] or a spot of bad timing on Capitol Hill, or whether something more serious was involved.

Perhaps most notable in the entire episode, which in a way was the most revealing series of events in President Carter's administration, was the explicit expression by the President of his underlying assumption that the entire intermediate layer that intervenes between a President and the grass roots of the country is fundamentally dispensable. How could a President not believe that a mock execution of his entire first team would be subjected to interpretations outside his own capacity to control? Or that the clientele both of affected government agencies and of the Democratic party might not readily accept his after-the-fact reassurances that the crisis was all for the best?

It is certainly true, as Mr. Carter discovered, that a President can go on television with some regularity and even some effectiveness. But others go on television as well: news commentators, for example, who frequently talk with their sources and their friends in that intermediate layer before broadcasting their views. Congressmen, labor leaders, ethnic and communal leaders, all speak with their own voices to their own constituents. When a President is fortunate, he shares these constituents amicably with allies in his party and with associated interest groups. But only briefly and occasionally can he override the longer-term loyalties of constituents to his own benefit.

Once again, the lessons of the nomination process proved mistaken. While it was necessary to go directly to primary electorates to capture the nomination, it was imprudent to infer from success at this endeavor that direct appeals for popular support would succeed without help in a more long-term relationship. Mass persuasion over the short run of a primary election season has a chance of succeeding without elite assistance; over the longer run, however, of a presidential term of office, successful mass persuasion entails successful elite persuasion.

Toward the end of his term, President Carter gave indications that he was beginning to accept the interest group basis of the Democratic party and hence to reward the coalition that had elected him, not merely the coterie that nominated him. This change of tactics, if not of heart, could be seen, for example, in the appointments to the cabinet of a southern mayor and a western mayor, Moon Landrieu of New Orleans and Neal Goldschmidt of Portland, and of a businessman prominent in Jewish communal affairs, Philip Klutznick. Likewise, he replaced his counsel on the White House staff, Atlantan Robert Lipshutz, with Lloyd Cutler, a well-known Washington lawyer, on whom he learned to rely. But the 1980 election returns, with their accompanying public opinion analyses indicating that the major reason for the massive Democratic defection was dissatisfaction with Jimmy Carter, suggest that in finally focusing on electoral rather than nomination politics Mr. Carter woke up too late.[105] His early concession of defeat on election night, wilfully jeopardizing his running mates farther down on the ticket throughout the western states, suggested that, however well he might have learned to cooperate with his fellow Democrats, it was uncongenial to him to do so.[106]

6. Conclusion

Jimmy Carter's pathway to the Presidency was originally dictated by the rules of the game as constrained, in particular, by the reforms of 1968–72. His response was to exploit the strategic imperatives these reforms brought into being. He built a personal following and invested most of his resources in states selecting their delegates early: Iowa and New Hampshire. In both states he succeeded not in forming a broad coalition but in mobilizing a faction, emerging first in rank-order among the numerous presidential candidates who had put themselves forward.[107]

The favorable publicity generated by these early results made plausible Carter's argument that he was the logical southern alternative to George Wallace in Florida. And so by the time he met with reverses in Massachusetts a week or so later the Carter bandwagon was rolling. By the time in June that Mayor Daley was ready to put his imprimatur on the Carter candidacy, Carter had won far fewer than a majority of delegates selected up to that point: his actual total was around 38 percent.[108] Though the traditional Democratic electoral coalition voted for their largely unknown candidate, Carter entered the White House as essentially a factional, not a coalition-based President. General awareness of the factional character of President Carter's outlook was to a degree retarded by the fact that Carter was not an extremist but rather a centrist in the ideological spectrum of the Democratic party. That he was well placed to embrace a broad coalition, however, did not guarantee that he would actually do so in forming his government, and he did not do so until his campaign for reelection was nearly upon him. No doubt this suited his personal style or character. More to the point, however, was the fact that the institutional constraints upon his personal preferences were weak. It is to be expected that a public official's private preferences and characterological traits will loom particularly large when no organizational constraints, no institutional cues, are offered. And this is the nub of the problem. Nothing in Mr. Carter's prior experience as a politician, certainly nothing in his experience of the nomination process, led him to the view that he needed to come to terms with the rest of the Democratic party. Greatly overriding considerations of party unity for Carter were concerns of good government, of addressing the policy issues of the day, of finding technically sound, comprehensive solutions to problems. These in turn could be presented to Congress and sold on their merits if not to Congress directly then over the heads of Congress to the "American people".

The people, however, were not so easy to mobilize in his behalf. In part this was a matter of bad luck: during the election year of 1980 the immediate rally-round-the-flag popular response of the early days of the Iranian hostage crisis decayed into a long-drawn-out period of frustration and seeming impotence. In part, no doubt, the two-step flow of communication from all the Washington-based intermediaries of the interest groups out across the country and from Congressmen to their constituents played a role in eventually diminishing President Carter's popularity. And finally, a certain pessimism and even grimness crept into Mr. Carter's own televised speeches, for example, on the numerous occasions when he addressed the nation's long-range prospects for the importation of inexpensive oil. In light of all the bad news that Mr. Carter delivered in person, over television, to the American people, it seems plausible to view the negativity of popular response as a triumph, rather than a failure, of presidential communication. Mr. Carter, not unlike President Nixon before him, evidently came to believe that he was a beneficiary of a system of governing genuinely free of mediation processes in the ordinary sense. He appeared to believe that contemporary circumstances sharply reduced the need to exhibit traditional presidential concerns with coalition building in order to govern. While both these conclusions are far from foolish deductions from the facts of genuine change in the presidential nomination process, they nevertheless proved to be off the mark. The relations between President and people are still powerfully mediated, so it seems, although by somewhat different processes and agencies than before. Presidents likewise still need interest groups and must build coalitions among them to govern successfully. This is not to suggest, however, that processes of intermediation have escaped the consequences of political reform.

IV

Wider Consequences: Political Intermediation, Mobilization, Accountability

Reform is a process that involves more than the enactment of prohibitions and requirements. Requirements and prohibitions newly enacted in the course of reform are added to an ongoing corpus of customs and regulations, producing a pattern of incentives to which different political actors, though they may vary in their comprehension, in their vulnerabilities, and in their resources, can nevertheless be expected to respond. It is not always easy to anticipate the behavior of actors as they learn how to operate over the newly contoured terrain that reform creates. It is certain, however, that the intentions of reformers will not comprehensively determine their behavior. Thus the assessment of reforms in operation is bound to take into account at least some unanticipated activity. And sometimes the newly emergent pattern of incentives has dramatic consequences for the maintenance of the political system. We have been following two such clusters of reforms, having to do with the centralization of authority over delegate selection and the control of campaign finance.

Implicit both in reformers' conceptions of reform and in the conceptions of critics is the assumption that political actors can and do learn to change their behavior with fair rapidity in response to changed rules of the game. This applies not merely to that part of their behavior evoked by the desire to evade criminal penalties but also to strategic and tactical behavior, activities designed to take advantage of new rules and regulations and to avoid adverse impacts upon actors' plans and ambitions. Because the presidential election process is bound to be rich in implications for the ambitions of politicians, it does not seem unreasonable to wonder ahead of time how reforms will change the structure of political incentives. Yet it is apparent that in the wave of reform we have been examining only rudimentary thought was given to such matters, and that reforms emerged in a context rather inhospitable to reasoned reflection of any sort: the turmoil of 1968 and its aftermath, and the trauma of Watergate and the impeachment crisis.

Hindsight may prove to be no better, owing to the difficulties, in a complex and evolving political system, of sorting out long-run from short-run effects, manipulable from uncontrollable phenomena. Nevertheless, it seems worthwhile to make an attempt to trace the effects of party reform at least a short distance out into the political system at large. While it cannot be claimed that reforms of the party system are solely and directly responsible for widespread changes in the underlying conditions of American politics, neither are they unrelated to them. Indirectly and in interaction with more diffuse changes in political life, political reforms can be shown to contribute significantly to emerging patterns and trends of American politics that deserve close attention. Significant among these are trends affecting the institutions and practices that maintain links between the general populace and elected officials.

1. Trends in Political Intermediation

The idea of a political party as a coalition of interests and groups bound together by many sorts of ties, including the hope of electing a President, is fast becoming an anachronism. Party is increasingly a label for masses of individual voters who pick among various candidates in primary elections as they would among any alternatives marketed by the mass media. Achieving financial support through mass mailings and through the public purse has displaced in importance the mobilizing of well-heeled backers and the seeking of alliances with territorially identifiable interest groups and state party organizations. The stimulation of coverage by the mass media and the building of personal organizations state by state through the activation of volunteer workers have been replacing dependence upon party regulars and state and local party leaders. Thus, interest group activity has been changing rather than diminishing. While interest groups that are mobilized in traditional ways around the economic interests or the communal ties of their members have declined in their political influence, the fortunes of other interest groups have been greatly enhanced because the managers of the mass media have decided to smile upon them. The decision to smile reflects world views common not only among news media personnel but also among the larger groupings of well-educated persons to which they belong, and while the political consequences are unmistakable, the criteria that make for success under these changed circumstances are not easily classifiable as partisan in their content.[1] Prominent among approved interest groups are those embodying what the media of mass communications accredit as disinterested rectitude, such as Common Cause and the Ralph Nader organizations, and those speaking for interests widely perceived as historically disadvantaged such as black, Hispanic-American, and militant women's groups. These groups, frequently and, for the most part erroneously, billed as grass-roots organizations, have taken on a new weight in American politics, in some cases achieving special recognition for their clientele in party rules and in law.[2]

I do not wish to argue that groups of this sort are especially worthy or unworthy as compared with state party organizations, labor unions, farmers' groups, associations of businessmen, or other interest groups organized on traditional lines around the economic or status needs of their clientele. But many of these newer groups—the ethnically based groups are the main exceptions—come into being and are sustained by different means than traditionally organized intermediate groups. These new groups are to an unprecedented degree the creatures of the media of mass communications in that it is their power to command news coverage and to be taken seriously by the news media that in some cases brought them into prominence and in all cases sustains their political influence.[3]

For good or ill a political system having intermediation processes heavily reliant upon the mass media is the sort of political system that is emerging in the United States. This does not imply that the media are reaching directly into the homes of totally atomized individual voters. Rather, political leaders are broadcasting to publics mobilized and organized around certain principles of attentiveness and inattention, and this has implications for interest groups and their success in politics.

Two recent political events in which modern mediation processes played a part may suggest a little of how they work. One is the non-selection of John Dunlop as secretary of labor by President-elect Carter in 1976. Another is the dismissal by presidential candidate George McGovern of Senator Thomas Eagleton as vice presidential candidate on the Democratic ticket in 1972.

It was widely assumed that Dunlop had a good chance to be Jimmy Carter's secretary of labor since, like another Carter adviser, James Schlesinger, his prior association with the Ford administration had ended in an honorable departure in which his personal integrity and political astuteness had to some degree been vindicated.[4] Unlike Schlesinger, Dunlop was a Democrat. Moreover, Dunlop was known to be far and away the favored choice of George Meany and the dominant faction of organized labor for the job as their ambassador to the new administration.[5] In indicating this preference, Meany had given due regard to a norm that prescribes a measure of independence from the labor movement for the secretary. Dunlop was not himself a labor leader but a college professor and dean, an academic specialist on labor relations with a long record of practical experience as an impartial arbitrator of labor disputes.

It may well be that President Carter never intended Dunlop to have the job, but it is at least intriguing to note that the appointment was publicly opposed in only one quarter: by leaders of black and militant women's groups who believed, whether rightly or wrongly, that in his prior government service and as Dean of Arts and Sciences at Harvard, Dunlop had been unsympathetic to their aspirations.[6] One interpretation of the course of events is that in a straight fight between one set of interests, labor, and another, black groups and the women's movement, over who was to be secretary of labor, it was not the traditionally organized group that won, even though the fight was over who was to be their point of contact in an administration run by the party they traditionally favored. It is not recorded that labor exercised a comparable veto over the appointments of black citizens and women in the Carter administration. The appointments of black and female Americans to administration positions were extensively monitored in the press, however, which in due course came to refer to these two groups, with the addition of Spanish Americans and Native Americans, as virtually the only interest groups whose progress was worth tracking.[7] This reflects news decisions made by independent news specialists, but these decisions are not empty of political consequences since they serve to confer legitimacy on some groups and their political claims and to withdraw it from others, as has been happening, for example, in the case of American Jews.[8]

Another facet of contemporary political intermediation is revealed by the Eagleton case.[9] Once it was discovered by the mass media during the presidential campaign of 1972 that Senator Eagleton, the Democratic vice-presidential nominee, had not disclosed episodes of hospitalization for severe depression, 't was only a matter of time before his senior colleague, Senator McGovern, had to remove him from the Democratic ticket. Why was this a foregone conclusion? Interested readers can search in vain through responsible journals of news and opinion for a serious discussion of hospitalization and recovery from mental illness as a disqualification for public office. Nor will they find a comparison of Eagleton's life history with that of his counter-part, the Republican vice-presidential nominee, Spiro T. Agnew, who had never been institutionalized or diagnosed as mentally distressed. In fact the news media rapidly reached a consensus that Eagleton had to go, but so far as an outsider could tell, it was not based on a discussion of the merits of the case so much as on the chagrin of media people at Eagleton's lack of candor with them and with Senator McGovern.[10] Gary Hart, McGovern's campaign director, wrote: "There were three possible bases for a decision on Eagleton's fate—personal, political, or factual. The decision to keep Eagleton was made on personal grounds; the decision to remove him, on political grounds. Both were wrong. The records, the medical evidence, should have been the basis for a decision in either case."[11] George McGovern, a nominee whose success depended not on the building blocks of interest group alliances within the Democratic party, but on the support of the news media, had no real choice but to dismiss Eagleton once the media consensus developed. McGovern's handling of the entire episode entailed sending messages to Eagleton by dropping hints among the journalists covering his campaign, and Eagleton was constrained to communicate back to McGovern in much the same fashion, a mode of communication which evidently greatly puzzled the journalists involved.[12]

Here we can note that one characteristic of the emerging style of political intermediation is that it is done in the sunshine. It is easy enough for anyone who has the price of a newspaper or access to the televised evening news to see what advice a political leader is getting from the news media. One difficulty with such an arrangement is that when politicians must announce themselves and their preferences on national television, they may get locked into position before they come to understand one another's point of view. Deliberation and negotiation, in which mutual accommodation and mutual learning are encouraged, are hard to arrange without causing one or more public figures embarrassment. Incidents of embarrassment in the public record are commonly thought to be an overwhelming disadvantage at election time. And so, participants are tempted into confrontation politics and moralism in order to look good. Of course in looking good, they may to an unaccustomed degree actually be good. But there are dangers here as well as opportunities.

These dangers are clear enough: interest-group alliances and alliances with state and local party leaders can to a certain extent protect a presidential candidate against unfavorable opinions in the mass media; without these alliances a candidate or a President has no court of appeal from media disapproval and this may seriously constrain the choices available to him. To be sure, interest group and party alliances do not arise without cost or without effort. As the case of the Carter Presidency suggested, intermediation by political leaders and interest groups more proximate to voters than the President may have a long-run impact that can be adverse to a President. But there are opportunities here as well as dangers, in part because establishing alliances with cooperative intermediaries constitutes a long-run strategy for enlisting that portion of public opinion subject to their influence.

Thus while there has been a shift in intermediation processes, toward more public and less overtly partisan vehicles of communication, it is not the case that modern forms of political intermediation free political leaders from important constraints. By trading labor leaders for television news commentators as intermediaries, Presidents and presidential candidates may have achieved a kind of freedom but only at a price. The Dunlop and Eagleton cases suggest what sort of price is involved. In a large-scale society based upon appeals from leaders to followers for their votes, it is evidently inescapable that some sort of division of labor will take place in which people specialize in working for mediating institutions that are separate from leaders and followers and undertake to link them. What sort of institutions these are, what sorts of values infuse their management, what sorts of messages they spread and re-

tard are bound to have political consequences and will influence the relations between leaders and their publics.

A sizable number of conventional notions exist about the
purposes and functions of political intermediaries in large-scale
political systems.[13] Intermediary groups are supposed to interpret the desires of ordinary people to leaders and to inform
publics of alternatives available to them, thus tutoring their expectations about the activities of government. Intermediaries
recruit and train leaders for politics, identify social problems
and suggest solutions. And they conduct long-range political
education, helping to form the loyalties of citizens toward the
state and providing legitimacy, a provisional sort of acquies-
cence to the underlying political order, upon which leaders of
the state can rely. Intermediary organizations teach political ob-
ligations to citizens and inform citizens of their political rights.

Traditional agents of political intermediation thus include the
nuclear family, the household, and the extended family,[14] the
school, primary communal groups such as exist in a work place
or church or neighborhood, or a voluntary association organized for the purpose of promoting some shared interest.

It has been apparent for some time that various trends in
modern society have eroded the monopoly that face-to-face and
geographically localized institutions once held over the time,
the attentiveness, and the loyalty of Americans. And so it should
come as no surprise to learn of their decline as monopolists of
political intermediation. The private automobile, the comprehensive network of roads, and the telephone have expanded
the potential for individual communication enormously and
have consequently attenuated the tyranny of geographical pro-
pinquity in determining the options that people have in adopt-
ing one or more organizations as their political intermediaries
of choice.[15]

Television has had a somewhat different effect. Since it is not
an interactive medium it has greatly increased the power of the
few who are at the focus of its attention and assured that certain sorts of standardized information are readily available on

a virtually universal basis. Political representation, another form of intermediation, has more and more become a specialized, even a professionalized activity.[16] Finally, the rules of politics are being rewritten to reflect all these trends and in some cases to facilitate them. National party conventions, for example, once were conducted primarily for the purpose of discovering what the delegates to them—organized primarily by their state party leaders—wanted to do. Today conventions are more frequently run according to scripts worked out by television consultants so as to maximize their advertising value to the viewing public.

Intermediaries are now heavily engaged in the various crafts of persuasion by means of mass media. In this situation direct and material interests are bound to become relatively less important as compared with symbolic interests.[17] Hence interest groups specializing in representing direct and material interests lose ground to those groups concerned with symbolic interests.

Political parties are losing whatever favored position they may once have had as prime repositories of the symbolic political loyalties of American voters. To say that political parties are in trouble is mostly to say little more than that state and local party elites have lost influence over some of the processes most important to their collective life, such as the making of political— especially presidential—nominations. Many observers claim that there has been a great burgeoning in the influence of so called single-issue interest groups, but what they are witnessing is not the proliferation of single-issue interests—which have always existed and attempted to influence the political process[18]—but rather a precipitous decline in the capacity of party elites through the control of their own institutional practices to resist, channel, accommodate, or limit the demands of these groups for extraordinary influence over the presidential nomination process.

Thus of the three major vehicles of political intermediation available to Americans—interest groups, the mass media, and political parties—all have been influenced by changes in the technology of communications available to Americans. Interest groups and parties have been significantly influenced by the mass media, and those groups and interests most dependent upon the maintenance of the structural and organizational integrity of interest groups and the parties have been the most disadvantaged by these changes.

2. The Political Mobilization of Citizens

The diminution of party loyalty in the electorate and the decline of turnout in presidential elections, insofar as these are bona fide phenomena of contemporary American politics,[19] are occurring incrementally rather than suddenly and massively. They may well be the results of changes at the elite level rather than, as reformers frequently argue, reasons for these changes.[20]

It is possible to speculate that changes in the party system might contribute in at least two ways to the modest declines observable in the political participation and party affiliations of ordinary citizens. First, in the course of discouraging coalition-building and encouraging factionalist strategies by presidential candidates, ordinary citizens may in greater numbers feel themselves and their opinions less taken account of by politicians. This phenomenon is frequently conceptualized as a distaste for extremist leadership, but the unpopularity of extremism may simply be a special case of the phenomenon of a factionalist strategy leading to the exclusion of large numbers of voters.[21] A factionalist from the center of the ideological spectrum no doubt may appeal to a larger share of the populace than an extremist; but competition for the first place votes of that larger share may be very intense, leading to tactics in which candidates seek to differentiate themselves and stake out exclusive first-choice territory. While this may capture the allegiance of a significant number of voters—presumably in the case of at least one candidate enough ultimately to win—it leaves the problem of the allegiance of those voters not attracted to the winner.

Coalition building explicitly seeks to deal with this phenomenon, and to bring large numbers of voters in under one tent. The argument here would be that neglect of this activity in the nomination process may be showing up in the mild but persistent increases in disaffection recorded in innumerable public opinion surveys and standardized measures of political participation.[22]

It is, surely, a paradox of great interest that as the education level of American adults drifts upward, at least some measures of political participation and affiliation, which at the individual level correlate positively and strongly with education, are for the general populace drifting downward.[23] In those presidential election years in which one of the major political parties nominates a political extremist there seems to be no difficulty in interpreting the result as abnormal defection and withdrawal from political activity of more moderate adherents of the party in which the blunder occurs. The increased necessity for political leaders to rely on factionalist, exclusionist strategies may simply be creating a permanent, chronic form of this same phenomenon, visible in its acute manifestation when extremists are chosen in the presidential nomination lottery which pits the leaders of many factions within a single party against one another.

A second process which may also be contributing to the chronic occurrence of depressed political participation and mobilization among ordinary Americans is the steady replacement of face-to-face, primary and geographically proximate interest groups with distant, symbolic and noninteractive mediation mechanisms. Such mechanisms may be reasonably efficient as devices for informing citizens of the date of an election and of some salient facts about candidates, but they are bound to be less comprehensively engaging to at least some ordinary voters than more personalized organizational structures.

The erosion of traditional intermediaries thus may in the long run contribute to the decay of political legitimacy within the constitutional order. This may be reflected in the palpable growth of disaffection with Presidents and in the phenomenon

frequently complained of—mostly by neoconservative observers—and labeled "ungovernability," a pervasive inability of political leaders to satisfy the expectations of voters.[24]

Political reform is frequently defended as a means for de-
creasing the alienation of voters—and the party reforms we have
discussed certainly were so described by their advocates.[25] Not
enough work has been done, however, to exclude the possibility that the arrow of causation actually runs the other way: by
contributing to the decommissioning of primary groups and
geographically based interest groups as prime political intermediaries and by encouraging factionalist, exclusionist political
tactics, reforms may be spreading the very disease that it is
claimed they are curing.

3. The Subculture of the News Media

If party elites are losing influence to media elites, then it seems
sensible to inquire in more detail about the sorts of performance we can expect from media elites in their new character
as leading intermediaries in the American political system's
presidential nomination process. This requires knowledge of the
workways of the news media, many of the essentials of which
have been carefully studied and are already well known.[26]

The fundamental axioms that underpin these workways are
reducible to two organizational imperatives: professionalism and
competitiveness. News media elites strive for professionalism,
which entails establishing their own account of day-to-day reality independent from that of the politicians whom they cover.
Yet of course news media reality must intersect at many points
with the realities perceived by political actors and it should come
as no surprise that there are frequent disagreements between
the two populations. News media professionals generally per-
ceive these disagreements as an earnest of their own professionalism and of their integrity.

Politicians occasionally take the view that journalists are attempting to create an adversary culture and consider news me-

dia people purveyors of radical chic.[27] The news media are better understood as chic than necessarily as radical. That is, news professionals are highly permeable to currents of thought among people like themselves, an increasingly educated and articulate segment of the population.[28] News professionals tend to act as transmitters and amplifiers for such ideas, especially ideas different from and even antagonistic to what governments are saying, but not because media people subscribe to some systematic anti-government political philosophy. Rather this antagonism is interpreted by news media elites as validating their independent professionalism.

Competitiveness acts in three ways to modify this picture, providing incentives for compression, for impact, and for convergence. Compression is a response to the severe time and space constraints that prevent news media managers from dealing with the vast bulk of the material generated by the news-gathering process by any but the most ritualized means. Competitiveness presses journalists to get a story and report it as quickly as possible. There is rarely time for reporters or editors to mull over what is going on long enough for second thoughts to have an impact on the way most stories are reported or played. Thus news stories tend to become highly stereotyped in the ways in which they are perceived by media professionals; it is a part of their job to shape stories into standardized dramatic structures. Paul Weaver has described one example—the horserace scenario of the presidential primaries—and Thomas Patterson has taken the trouble to document the way in which this scenario totally saturated the coverage of the 1976 pre-convention campaign.[29] Other standardized scenarios: The man-bites-dog, the little guy takes on system, the two politicians who trade insults are easily recognized fare to experienced consumers of the media.

To my knowledge nobody has done a census of the most frequent dramatic plots that occur in the news, but it seems reasonable to suppose that they are severely limited in number.[30] The most important of these in the context of primary elections is no doubt the deviation from expectations: "wins" and "losses" by various candidates are evaluated by news professionals in the light of how well they "expected" to do, or were expected to do by some consensus of observers. In some primary elections, "Expected" becomes a phantom candidate who does not always obey the laws of logic or mathematics: Johnson did not do as well as "Expected" in New Hampshire in 1968; McCarthy did better than "Expected." The fact that Johnson outpolled McCarthy was not given anywhere near the coverage as was their success against the phantom candidate. Because expectations are so significant in shaping the plots of primary election stories, journalists take some pains to establish expectations in advance. This entails attempts to get candidates or their managers to specify how well they hope or expect to do, any deviation from which creates a news peg. Candidates and their managers, needing news coverage for publicity purposes, must maintain rapport with journalists, but know that it is in their interest not to specify a number that will stimulate a story reflecting disappointed expectations. Such stories dry up contributions and cause candidacies to falter.

Competitiveness also encourages attempts to maximize impact, both on readers and on political actors. The need for impact on readers and viewers raises the issue of sensationalism, about which journalists agonize a great deal. Competitiveness, that is, the need to compete for the wavering attention of mass audiences, and thereby to sell the news product as a successful advertising medium, clearly creates an organizational bias in favor of highlighting the dramatic, the emotionally overwrought, the bizarre, and the atypical. Professionalism as a norm no doubt acts to constrain this bias, but as examples such as the 1981 Pulitzer Prize scandal involving the *Washington Post* suggest, the temptations to sensationalize are ever-present.[31] Or another example: It is well known that there are usually more deaths on the UPI ticker's first account of any disaster, a harmless attempt, it is well understood, by the smaller wire service to nudge the better established AP's story out of as many papers as possible.[32] Discussing the coverage of presidential campaigns, NBC broadcaster Roger Mudd says:

> Over the last 15 years, as competition has sharpened between the networks, none of us is content to let an event be an event. We have to fix it. We have to foreshorten the conclusions, hasten the end, predict before anyone else does who's going to win. We have to take an issue on our own terms and we won't let the candidate lay out the issues on his terms.[33]

Impact on political actors increases the credibility of a news organization to its customers, hence helps it compete.[34] American television viewers are used to the phenomenon of the most significant question at any presidential press conference according to CBS News being asked by the CBS reporter and so forth. Modern Presidents must be prepared to announce their intention to ask Congress to declare the same war three separate times—once for each network—if they intend to do so in response to a press conference question.

In order to be convincing purveyors of reality, if possible more convincing than the competition, journalists must get as close as they can to the sources of events. This means access to the political leaders to whom they give publicity both for themselves and their views as a quid pro quo for the proximity that lends verisimilitude to the journalists' accounts.[35]

Competitiveness thus entails snuggling up to news sources and works at cross purposes with the imperative of professionalism to maintain independence. The tensions, both institutional and personal, that result from this contradictory set of demands are endless. An elaborate code of ground rules—frequently misunderstood and violated—has grown up in Washington governing the rights of journalists to publish information they come across on social occasions, at off the record encounters, at "background" briefings, and so forth.[36]

One would think that competitiveness would lead to a strong push toward product differentiation, but this depends also upon marketing strategies. On the whole it does not work that way among the most important national news organizations.[37] Appealing to the fat part of the market, which is the marketing strategy of the major suppliers of news, rather than to specialists, entails a systematic suppression of detail and a search for instantaneous coherence, however spurious, that causes reporters and editors alike to monitor and tacitly or explicitly to collude with the competition.[38] Nobody who intends to supply mass publics with their daily ration of news can afford to be out on a limb too often, peddling what may come to be viewed as an idiosyncratic version of reality. Since the realities for a newspaper are mostly social and political realities, they depend on consensual definitions of the situation, and most news organizations most of the time willingly participate in that consensus, sometimes greatly to the advantage of undeserving politicians, sometimes greatly to their equally unmerited disadvantage.[39]

The two principles of professionalism and competitiveness thus account in a sketchy way for most if not all of the evils of the press as interpreted by critics as diverse as Spiro Agnew[40] and Daniel Patrick Moynihan, without recourse to imputations of reportorial bad faith, ugly motives, or malicious intent.[41] This is not to deny that any of these exist but only that there are better, more plausible ways to account for a larger proportion of news coverage by presuming that the product we see is the result of conventional mass marketing strategies under highly competitive circumstances and widely shared subcultural norms of professional conduct.

4. Governing Without Parties: Some Possible Results

Let us imagine ourselves in a world that looks a great deal like the United States but where more or less representative elites carry less of the burden of governing, intermediation is less a face-to-face and more a mass-marketing phenomenon and there is a greater reliance upon referenda, plebiscites, and other sorts of voting and manifestations of public opinion. These mechanisms can be characterized as enabling traditional forms of intermediation to be bypassed and hence as conducive to "direct" democracy, although, as I have tried to indicate in the discussion of the mass media as agents of intermediation, the directness of direct democracy in a very large scale society seems to me illusory. What are the consequences for government that might follow?

(1) *Crazes or manias.* We might expect an increase in the amplification of the intensity of short-term trends of opinion. Political leaders may come to believe that certain behavior is strongly demanded of them, even though these demands in fact have only weak or transitory support in the populace. In the 1950s for example it was believed that the U.S. population supported Senator Joseph McCarthy to a much greater degree than they actually did.[42] Elites on that occasion tended to interpret the movement of small straws as evidence of a big wind.[43] Their capacity to monitor short-term trends today is greatly enhanced by the growth of the use of public opinion surveys as a substitute for listening to interest group demands as formulated by their leaders and by their lobbyists. President Reagan is able, for example, to select his legislative priorities on the basis of public opinion polls made available to him on a weekly basis.[44] This is little different from President Carter's access to public opinion surveys, or his inclination to attend to them.[45]

(2) *Fads or social contagion.* Opportunities will arise for the geographic spread of sentiments, both real and imputed, from areas where they exist to areas where they do not. The enthusiastic, though uneven, epidemic of attempts to put tax limitation referenda on ballots across the country in the wake of the success of California's Proposition 13 can be interpreted in this light.[46] Real estate prices did not skyrocket everywhere in the presence of multibillion-dollar state government surpluses. However, there is no Mann Act for ideas forbidding the transportation of immature notions for immoral purposes across state lines. Rather there is the tendency for issue entrepreneurs to press their luck in one constituency after another, rather like burglars rattling all the doorknobs in the neighborhood look-

ing for weak defenses. Modest turnouts in a referendum can constitute just such a weakness.[47]

(3) *The resuscitation of ideology.* I interpret the term ideology broadly as meaning doctrines that elites invoke to capture the attention and induce the compliance of mass publics.[48] Presumably in a mass persuasion system elite politics consist in part of fights for control over the right authoritatively to interpret equivocal or incoherent events. Since most events in their uninterpreted state are somewhat ambiguous in their meaning, this is no small matter. For example, an incident of urban violence is generally an incoherent event to those at a distance from it. Does it mean that there is too much racism in society or that there are not enough police in the slums? Are looters expressing legitimate social grievances, or should they be shot? Such questions do not often have to be resolved authoritatively and incorporated into an overarching world-view in political systems dominated by representative elites. However, they must be interpreted and a social definition of such events agreed upon when more transient expressions of public opinion are the main source from which political elites must draw their entitlement to govern.

(4) *This underscores that elites do not disappear or become less significant in governing under a mass persuasion system.* They are, however, less accountable to one another and more subject to the constraints of popular fashion. They must learn to feed the mass media successfully, to cultivate different virtues, e.g. less patience of the sort employed over the bulk of his career by Richard Daley,[49] more indignation à la Ralph Nader.[50] Interest groups that organize themselves around such anachronisms as state and local party systems are bound to lose out to those that are skilled in packaging their ideas in ways that appeal to reporters and news media gatekeepers.

(5) *Heroes and bums.* Such a system runs on name recognition, on celebrity, and on typecasting. An odd similarity between elite persuasion and mass persuasion systems is the inside track both give to children of the prominent who want to

pursue political careers. Different mechanisms of recruitment are involved, however. In a mass system, name recognition is of great significance in the competition to overcome the inattentiveness of mass participants. Politicians must put enormous effort into structuring the ways in which they are presented by and to the news media, since getting caught on the wrong side of a too-visible political issue may typecast a politician and affix a black hat on him for a long time. Typecasting is, in any event, ubiquitous. Henry Jackson and Morris Udall, both liberal Democrats according to the records they have compiled over the 18 years they have overlapped in Congress, were located respectively on the far right and the far left among the Democratic presidential hopefuls of 1976.[51] This reflected dramaturgical necessities of mass persuasion politics and had only a little to do with their approach to political issues. Nevertheless it was necessary for them to collude in the scenario because their political situation demanded that they differentiate themselves and cultivate their respective factions rather than attempt to build a majority coalition.

The wider consequences of changes of this character for the actual workings of government are difficult to fathom. Two speculations at least are worth considering and both suggest an increase in certain sorts of stress in the political system.

The forces that have transformed the presidential nomination process have been far slower to affect the thousand-odd nomination processes that supply Americans with their candidates for the House and the Senate. This continuity at the level of nomination politics persists even though Congress is a greatly changed place from the way it was even a decade ago.[52] The previously moribund House Democratic caucus has become intermittently available as an instrument of majority will when that will exists. But the mobilization of that will is by no means the unchallenged prerogative of a Democratic President. Republicans, with their far narrower ideological spectrum, may find themselves better able to coordinate between President and Congress when both institutions are controlled by Republicans, but of course when a Republican President faces Democratic congressional control the mobilization of a cross-party coalition will be necessary. Congress is today better equipped than ever to find its own way without presidential guidance or cooperation into the intricacies of policy and to arrive at its own balancing of forces and priorities.

So long as elections in America are of the staggered, prescheduled, non-referendum type, and so long as nominations to Congress are decentralized, it will be difficult if not impossible for a President to attempt on a regular basis to persuade Congress by going over its head to the people. This is no substitute for the hard, frustrating, and frequently unavailing work of doing business directly with Congress. Because of the divergence that currently exists in the ways in which the President and Congress mobilize their electorates and arrive in office, we may be entering an era in which tensions and misunderstandings increase between the President and Congress, even when both are controlled by what is labeled as the same party. When one party controls the Presidency and another one or both houses of Congress, the need for carefully crafted diplomatic relations between the two branches is not lessened.

A second possible consequence of governing without the intermediation of parties and traditionally organized interest groups is a decrease in the reliance of political executives on the accumulated expertise of the permanent government. When the top of the government is dominated by client-centered political executives and executives recruited from party-oriented interest groups, bureaucrats supply technical information, policy analysis, know-how, and knowledge of programs. Other sorts of political executives erode the division of labor in which leaders of the permanent government have a distinctive place. Some cabinet officers—possibly symbolic appointees or presidential ambassadors—are liable to ignore the agencies they head altogether and will take orders with varying degrees of diligence from the White House. Others, most likely specialists and Washington careerists, may undertake policy evaluation and analysis in competition with their own agencies. Much of this activity no doubt will result in vastly improved understanding at the top of government of policy alternatives and their consequences. Moreover widespread ad hoc policy analysis at the political executive level can create over the long term a sizable number of experts outside the government who can contribute to the understanding of policy, to the consideration and selection of technically ingenious alternatives, and to ultimate public acceptance of otherwise inscrutable policy choices.

But as the comparative advantage for policy-making of being inside a government bureaucracy begins to diminish, the caliber of the agency is endangered. As agencies begin to view their nominal superiors as competitors in the provision of what they supply best—expertise—the incentives to firm up other sorts of alliances, with clientele or with Congress, increase. So, also, do incentives to depart the government altogether.[53] Thus over the long run without strenuous effort directed toward the cultivation of good will between political executives and career bureaucrats, the growth of knowledge among political executives may actually diminish the amount of knowledge available within the government.[54]

The purpose in elaborating these hypotheses is not to make the claim that these sorts of outcomes are the natural and direct result of party reform. Rather, they are the possible results of other forces working in the political system when parties fail as mechanisms of intermediation and other devices fill in the spaces left by the shrinkage of parties. If parties do not mediate between President and Congress, if parties do not provide the basis for the formation of a governing coalition that orchestrates the activities of the executive branch, if parties do less in the way of organizing mass-elite relations, what alternatives are readily available to take up the slack? These are the questions I have been addressing. The chain of causation implied in this discussion is that party reform has weakened parties, and in that sense has nourished alternatives to party. These alternatives—the mass media, interest groups favored by the news media, presidential candidates and Presidents only loosely attached to their parties—have grown in importance for a great variety of reasons, most of which are beyond the scope of the present inquiry. Thus our exploration of significant change in the body politic focuses not upon the proliferation of bacteria, many of which might well have flourished anyway, but on the weakened condition of the host, which has provided avenues of attack for functional alternatives to parties.

5. *Accountability*

An interesting problem of social control that emerges from these reflections has to do with establishing accountability to the enduring values of a democratic society in elites that gain power in systems of mass persuasion. In the system of multiple elites evolved under the U.S. Constitution, accountability has been maintained mostly by means of the checks and balances explicitly built into the machinery of government and by means of a party system that structures the alternatives for electoral choice. In such a system elites monitor one another. What sorts of mechanisms provide checks and balances for newly emerging centers of power that are taking the place of parties in the presidential nomination process, such as those which control the mass media? It is no doubt too early in the development of the new system of intermediation to give a definitive answer. One traditional response—that the mass-marketed news media monitor one another by virtue of their freedom to publish and their multiplicity—has some long-run merit. The needs of the American system of news manufacture and supply to arrive at shared and consensual views on many issues work against the effectiveness of this norm, however, and suggest that the news media are more effective in establishing certified views than in providing access to a diversity of voices. Diversity in publishing and broadcasting in any event implies differentiated, diverse, and small audiences and does not address the issue of access to the large audience of basic news consumers. The resistance of the major news media to such innocuous monitoring and corrective devices as independent, non-governmental press councils is well known, as is their disinclination to admit error. Finally, there is the problem of the increasing consolidation of ownership of newspapers. A more responsible consideration of this cluster of problems is long overdue.

To what extent do newer methods of intermediation perform the full range of social functions that earlier fell to the old? It is in this case also too early, or perhaps too difficult, to say. Clearly, political education continues apace. Demands continue to be made upon the political system in the name of citizens and blocs of citizens. The government seems to enjoy more than enough legitimacy to get its routine work accomplished without difficulty. If there are any nagging doubts about the newer forms of political intermediation, they surround the issue of the accountability of elites. But on this issue many traditionally organized political intermediaries are also vulnerable. To what extent did George Meany represent his rank and file? And how did he know? We can suspect, without really knowing, that on one great set of issues, civil rights, at least as it had to do with access to blue collar jobs and neighborhoods, labor union leadership was far in advance of the attitudes of their rank and file. Whether they were or not, the rather sluggish methods of accountability that have prevailed in the labor movement permitted labor leaders to lead on this issue more or less with impunity.

The accountability of the leaders of the newer interest groups, such as Common Cause or the Ralph Nader organizations, to the people in whose behalf they ostensibly speak, or even to their paid-up members may be even more tenuous. Indeed, of the eighty-three "public interest" groups studied by Jeffrey Berry, who seems to be the only scholar to have investigated the question, 30 percent had no members at all, and consisted only of lobbyists. Of the rest, Berry says flatly,

> If "democracy" is taken to indicate that constituents have a formal opportunity to select their leaders, . . . public interest groups are not democratic [but] quite oligarchic in nature.[55]

Fifty-seven percent (33 out of 58) of the public interest membership groups have no structure that ostensibly elicits and considers
membership opinion.[56]

The capacity of such groups to deliver the votes of the people they organize directly must at the least be highly questionable. In any event, the numbers involved are quite small.[57] The
real clients of new style interest groups are news media managers. So long as newspaper, magazine, and television people
believe in the rectitude of Ralph Nader, and others similarly
situated, and defer to them and their judgments, and give them
publicity, politicians will feel obliged to defer to them as well.
And leaders of these sorts of interest groups can continue to
claim to be authoritative interpreters of the desires of people
at the grass roots.

It will no doubt come as a disappointment to some observers
that current trends in political intermediation do not seem to
provide for greater accountability and hence greater openness
and democracy in some larger sense. In fact it can be argued
that groups equally deserving as the ones that have gained
ground have lost ground because of their propensity to organize along more traditional communal or face-to-face lines. In
particular the groups traditionally served by city machines, geographically compact and ethnically homogeneous neighborhood groups, for example, may well be suffering greatly.
Women with a history of activism in political parties may be at
a disadvantage compared with women with a history of activism
in the women's movement. This is a complicated business, leading as it eventually may to distinctions between nominal members of disadvantaged groups (e.g., women) and authorized
members (e.g., women active in the women's movement).[58] Life,
as Presidents increasingly have occasion to remind us, is unfair
and illustrations of that uncomfortable maxim can be found in
the field of political intermediation as readily as in any other.

The newer methods of intermediation operate unevenly in
the political system. They have far more to do with the elec-

tions of Presidents or governors of large states where mass publics can be efficiently mobilized and party leaders are relatively weak than with nominations and elections to Congress. They involve the manipulation of attitudes that may be tenaciously held as well as those that are more ephemeral, and they evoke loyalties that are probably mostly nontransferable. For these reasons one must sharply question whether it makes sense for any President to believe, as Jimmy Carter evidently did, that the type of intermediation that got him nominated and elected is adequate to the formation of a governing coalition.

For the purposes of governing, alliances with relevant factions and blocs in Congress and with national interest groups not necessarily organized on the new principles, are bound at a minimum to prove useful. Efforts to ignore, bypass, or run roughshod over such groups by appealing over their heads to the people, are doomed on at least two counts. First, the appeal to public opinion itself is likely to fail because of the fickleness and lability of mass public attitudes on most issues, and because of the nontransferability of a President's popularity, when the President is popular, to the objects of a President's desires. One week a President may be riding high in the news media because of a couple of stunning congressional victories; the next week an adverse reaction on Wall Street can jeopardize the rest of his program. Second, even if by some unusual combination of circumstances, public opinion does for once yield to a President's entreaties, the effects may or may not reach Congress or influence congressional disposition of an issue. Members of Congress, after all, have their own constituencies and their own means of reaching them, and they may find themselves ill-disposed toward a President who prefers to deal indirectly with them through what they may interpret as coercion rather than face to face and in a spirit of mutual accommodation.

It may be entirely feasible for a President to live according to the dictates of a strategy that is concerned exclusively with his own renomination and reelection if he wants nothing of Congress and nothing from traditional interest groups. But if a President means to do more than hold onto office, he will have to face the pockets of pluralism that remain in the political system and somehow deal with them. If Common Cause and other reformers succeed in their efforts to reduce the capacity of Congressmen to facilitate their own reelections by attenuating their privileged access to publics in their home districts,[59] it may well be true in some future time that Presidents will be able to get their way on matters of public policy by direct appeal to public opinion. At the moment, however, checks and balances preclude this under ordinary circumstances.

Thus traditional interest groups based on geography and economic self-interest still exercise meaningful influence in Congress. So, with a few exceptions, do state and local party organizations, in the shrinking number of places where these retain their traditional strength. In the presidential nomination process, however, these groups have lost ground to groups more approved of by the mass media and organized on different principles, where interest group leaders reach their own constituencies by means of media-based publicity rather than through some sort of internal structure of communications, whether communal or corporate. The enormous growth of the influence of the news media over the presidential nomination process has led Presidents and their professional political consultants more and more to believe that a high level of public approval is the central resource needed to govern and that the attainment and maintenance of high popularity is relatively unconstraining as compared with the constraints of bargaining with interest groups. It is not required that Presidents entertain such beliefs, but it is not necessary to give these beliefs up (when Presidents hold them) in order to succeed in manipulating the nomination process to their advantage. These systemic changes consequently were not merely to be found at the root of President Carter's difficulties in governing; difficulties of this sort will in time come to be seen less as idiosyncracies of the Carter administration than as realities to be confronted by any administration.

V

Political Reform
and Democratic Values

1. Primaries and Participation: Some Problems

The confusions of the historical moment that stimulated the reforms we have been considering do not permit a straightforward assessment of the goals and motives of those most responsible for their promotion. Certainly among the ideas animating the drive for reform, however, were thoughts that the presidential selection process could be improved by making it possible for more people to participate. Consequently a set of guidelines having as a proximate consequence the opening of delegate selection to a series of primary elections could be defended, at a minimum, as in some sense contributing to the democratization of presidential nominations: where previously only a handful of party leaders may have selected delegates to the national party convention, now millions of voters have a part in the process. In no case has the advent of primaries reduced the number of people involved. Observers have been ready on these grounds alone to argue that whatever else the party reforms may or may not have accomplished, they have surely opened up a process that was sorely in need of democratization.[1]

Table 5.1 Comparison of Democratic primary electorate and Democratic general electorate, 1976. (Percentages of voters)

State	Less than high school education		Black		Over age 65		College degree or beyond		Income over $20,000/yr.	
	Primary	General	Primary	General	Primary	General	Primary	General	Primary	General
California	11	27	12	15	9	19	34	17	35	23
Florida	13	30	8	14	23	28	28	17	26	16
Illinois	19	34	15	16	8	17	23	6	25	21
Indiana	21	40	10	11	7	24	13	6	17	11
Massachusetts	12	19	2	3	10	18	36	22	24	24
Michigan	20	30	11	22	11	17	20	10	16	17
New Hampshire	11	18	—	—	6	13	38	18	15	14
New Jersey	12	41	17	26	9	16	35	14	37	19
New York	15	31	15	20	15	14	32	19	32	16
Ohio	15	32	11	17	7	19	25	10	23	17
Oregon	19	18	n.a.	n.a.	23	13	23	20	15	23
Pennsylvania	17	32	8	15	9	15	23	15	20	9
Wisconsin	18	25	3	7	13	19	22	15	n.a.	n.a.

Note: n.a. signifies data are not available. Dash signifies less than one percent.

Source: Democratic Party, Commission on Presidential Nomination and Party Structure, "Openness, Participation and Party Building: Reforms for a Stronger Democratic Party," Washington, D.C., January 25, 1978, pp. 11–13.

Two venerable lines of argument in political science suggest that this accomplishment may be illusory. The first has to do with the sorts of participants who dominate primary electorates. The second has to do with the formal properties of choice in primary elections as they are run in the United States, in which outcomes are produced by aggregating the first—and only the first—preferences of voters who spread their selections over a broad field of contenders.

Because primary elections have long been a feature of American state politics, there is a considerable history to the argument establishing the first point. As V. O. Key, Jr., one of the ablest of those who have studied the matter says, "The American political tradition caps decisions made by popular vote with a resplendent halo of legitimacy." Yet the chapter of Key's book of which this is the first sentence is titled: "Participation in Primaries: The Illusion of Popular Rule."[2] Key shows that for the large number of state primaries which he examined, in several states and over a number of different elections, "the effective primary constituency" is often "a caricature of the entire party following."[3]

Key's major findings—that state primary electorates were unrepresentative of the state electorate, and that candidates frequently needed to mobilize only very small numbers of voters to win—are confirmed for more recent national primary elections, notably by work of Austin Ranney, James I. Lengle, and the Democratic party's own post-1976 rules-examining body, the Winograd Commission.[4] Table 5.1 gives Winograd Commission findings comparing selected characteristics of Democratic primary electorates with Democratic voters in the 1976 election for 13 states.

Consistently, and not only in the case of 1976, those voters lower down on the socioeconomic scale are disproportionately missing from primary electorates. Lengle's work shows in addition that socioeconomic differences are associated with ideological differences. As people gain in education they are predisposed to identify themselves as more ideologically extreme.

This suggests that a lack of demographic representativeness in a primary electorate may produce significantly different results in the types of candidates chosen to lead the party. Direct tests of this inference among California Democrats in 1968 and 1972 strongly sustain the notion that candidates are perceived differently and supported differently by voters differently situated along the socioeconomic spectrum. Similarly, differences occur in their views about the importance of different issues.[5]

The better-off voters who vote in primary elections are thus, in their hundreds of thousands, in a sense stuffing the ballot box much in the manner of the fans who vote for their favorite players in the election that from time to time determines who will play in organized baseball's all star game, as Richard Schier's inspired analogy illustrates:

> In 1970, baseball Commissioner Bowie Kuhn announced his intention "to give the game back to the fans" by permitting them to pick the players, this despite two earlier periods of experimentation with this method of selection. In 1957, spirited balloting by Cincinnati fans elected seven Reds to take the field along with Stan Musial as the starting lineup. (Managers chose the pitchers and reserves.) After this debacle player selection was restored to the players, managers and coaches until Commissioner Kuhn revived participatory democracy in 1970.

> The repetitive access to the balloting enjoyed by fans in Los Angeles, Philadelphia and Cincinnati—cities with huge ballpark attendance—gives them an advantage similar to that with which the news media's attention to the first primary endows the whims of New Hampshire voters. Neither in baseball nor in politics has the ideal of one-man, one-vote been realized. . . .

> Two years in a row, American League fans elected an outfield in which none of the three players were comfortable in center field. This same problem plagued the Democrats in 1972. . . .[6]

> Criticisms of these two processes are strikingly parallel: An activist minority can impose its choice on the more apathetic majority; good candidates may be overlooked or left out; certain localities occupy a strategic position that gives their residents a disproportionate influence; and the length of the process makes timing an element of great significance.

We have the word of a member of the McGovern commission on party structure and delegate selection that the new rules "were consciously designed to maximize participation by persons who are enthusiastic for a particular aspirant in the year of the convention." That they have succeeded is not in doubt as the nominations of both George McGovern and Jimmy Carter testify. But the experiences of these two gentlemen also suggest that this enthusiasm may wane quickly.[7]

In spite of the supposed democratization of the selection process, and the increase in the number of women and black delegates, those delegates actually chosen to be present at Democratic conventions do not appear to differ greatly in their socioeconomic status from delegates chosen by old party elites. As Haynes Johnson of the *Washington Post* said of the delegates to the 1972 Democratic convention that ratified the selection of George McGovern: "Whatever else is new in American politics this year, the old ingredients of money and education and class still dominate the process.[8] A *Post* questionnaire sampling about half the delegates turned up 39 percent as holders of postgraduate university degrees. In the general population about 4 percent had done some graduate work. Income figures were:

Table 5.2 Economic standing of delegates selected under new rules

Annual income	1972 Democratic delegates (Post survey)	Nationally (U.S. Census)
under $5,000	6%	18%
$5–9,999	10	32
$10–14,999	20	27
$15–25,000	31	18
over $25,000	31	5

Source: Haynes Johnson, "Portrait of the New Delegate," *Washington Post* (July 8, 1972)

What has changed is that candidates have gained and state party leaders have lost the right to designate who the delegates shall be. The numbers of designees available to the various candidates are increasingly prescribed by the results of primaries, which allocate each state's entitlement in different proportions to different candidates. In picking their delegates, candidates must adhere to whatever rules the national parties have imposed about the demographic characteristics of delegations. In practice this has meant that Democratic conventions have in-

Table 5.3   Socio-economic standing of 1972 Democratic minority delegates

Education (years):

	Hispanics (n=53)	Blacks (n=162)	American Indians (n=45)	For comparison: General Population
0–8	2.0%	1.2%	0.0%	24.8%
9–12	9.4	14.2	20.0	52.2
13–15	37.7	30.9	26.7	10.9
16+	50.9	53.7	53.3	12.1
	100.0	100.0	100.0	100.0

Income ($1000, per annum):

	Hispanics (n=58)	Blacks (n=176)	American Indians (n=50)	For comparison: General Population
less than 10	19.0%	25.1%	20.0%	43.6%
10–14.9	41.4	27.8	30.0	26.1
15 or more	39.6	47.1	50.0	30.3
	100.0	100.0	100.0	100.0

Occupation:

	Hispanics (n=54)	Blacks (n=165)	American Indians (n=50)	For comparison: General population
professional, technical	40.7%	52.7%	32.0%	8.8%
management	11.2	15.8	20.0	6.2
clerical	7.4	4.8	10.0	10.9
student	24.1	12.7	8.0	5.7
housewife	5.6	4.2	12.0	27.6
unemployed	1.9	1.8	4.0	3.7
other	9.1	8.0	14.0	37.1
	100.0	100.0	100.0	100.0

Source: *The 1972 Convention Delegate Study* (ICPSR #7287), made available by the Inter-University Consortium for Political and Social Research.

creasingly "represented" categories of people previously "underrepresented"—usually with high-status, well-educated, and well-off members of the groups in question, as Table 5.3 shows.

The second sort of distortion which primary elections promote is harder to describe. It rests on the seemingly intractable fact that once the number of alternatives available to an electorate rises above two, and so long as only the first choices of voters are counted, there is a nontrivial likelihood that the plurality winner of such an election will turn out to be unwanted by a majority. The process provides no method for achieving a more acceptable result.[9]

The problem to which this possibility points is one of finding a method for expressing accurately the widely spread sentiments of voters in the result of an election in which there can be only one winner. If out of all possible alternatives there is one choice that is overwhelmingly popular then of course this problem does not arise. Frequently, however, there is no such popular solution, and the voting population has a complex structure of preferences. Among the elements of complexity it is possible to identify such phenomena as cycling, or an inability to determine a majority choice.

No clear first choice of a majority emerges for any candidate when successive pair-wise comparisons yield the result of candidate A beating B, B beating C, and C beating A. In theory, this cycling phenomenon happens whenever the pattern of preferences among three or more voters is distributed among three or more candidates in a manner approximating the following:

Voter	1	2	3
Preference	A	B	C
Rank	B	C	A
	C	A	B

Given the large array of alternatives frequently present in a primary election and the large number of voters, it is easy to imagine a structure of preferences which would actually result in cycling if successive pair-wise comparisons were permitted to occur. In practice, however, primaries do not permit the full array of voters' preferences to be expressed. Only first choices are given. Not only does this risk failure to produce a winner actually preferred by a majority, but it also suppresses information which it is necessary for voters to have in order to adapt their own views to the spread of preferences, perhaps by voting strategically so as to concert on satisfactory even though not first-choice candidates. So in practice when the structure of preferences in the electorate is such that there is actually no spontaneous, clear first choice, primary elections conceal this fact without necessarily producing the same winner as would emerge if more information about voters' preferences were to be disclosed to those doing the choosing.

A significant implication is that when sentiment is spread among many possible choices the results of primary elections are quite likely to be far more responsive to formal properties of the choice-making situation—which of all possible pair-wise comparisons happens to be first in sequence, for example, or how many options are available to voters in a given primary election—than to the underlying structure of preferences in the voting population.

Even if aggregating the responses of large numbers of voters spread among large numbers of alternatives were not a problem, eliciting an accurate expression of individual voter preferences from first choices alone would be. First choices disclose only a minimum about which among the array of alternative candidates might be satisfactory and which unsatisfactory. A version of this issue has been called the "intensity" problem, in which it is proposed as a desideratum of a decision-making process to be able to register how strongly voters feel for or against a given candidate.[10] Many voters may feel more strongly about which among the possibilities presented to them are unsatisfactory than satisfactory. This can in principle be discovered by a device known as approval voting, in which voters are permitted to vote for as many candidates as they like—but only once for each candidate meeting their approval.[11] American primary elections do not employ approval voting, however.

Some voters may feel strongly about their first choice but indifferent thereafter. If passions run high but choices are spread around, it is likely that the ultimate winner will be unpopular and, depending on the confidence of voters in the choice mechanism that produced him, possibly even damaged in his overall acceptability. This is arguably one reason why sharply contested primary elections sometimes lead to losses in the general election.[12] My argument is simply that this can in part be a result of the mechanism of choice.

What is missing in the relatively automated processes of choice involved in the simple allocation of securely pledged delegates, allocated according to the results of a ballot in which first choices only are allowed, are devices for arranging compromises. Second choice candidates enjoying widespread approval are unable to get into a game in which only first choices are counted. This becomes a problem because not all first choice candidates of some voters are minimally acceptable to other voters—conceivably even to large numbers of voters to whom the party wants to appeal in the general election. Processes of deliberation ought in principle to be able to smooth out some of the difficulties that arise when a plurality first choice candidate causes divisions of this sort in the party electorate. Primary elections do not deal with the problem at all, leaving it to fester until the general election.[13]

In summary, then, it can be shown that a candidate selection process heavily reliant on primary elections fails to meet a number of tests that it would be desirable for this sort of decision-making process to meet:

(1) It cannot be relied upon to identify a candidate who would beat each of the others in pair-wise comparisons. Successive pair-wise comparisons are not, in general, what primaries do. More characteristic of early primaries is a large number of entries. Sometimes there is in fact no candidate who

would beat all the others in pair-wise comparison, but even when this is true the sequence of primaries and other formal properties of the system dictate the result, producing what in effect may well be the wrong winner.[14]

(2) Even if primaries produce a clear winner, as, stimulated by the needs of the news media for a coherent and clear result, they usually do, that winner may be less effective than some other satisfactory candidate might have been in winning a general election.

(3) Primaries do not register second choices, intensities of feeling, or anathemas, all valuable information about the structure of voters' preferences in forestalling premature closure on the wrong choice.

(4) If primaries succeed in reflecting the choices of any population, it is not a population fully or accurately reflective of the party electorate, or the general electorate.

V. O. Key says:

> The small size of the blocs of voters necessary to win nominations has a most significant consequence for the nature of the party . . . The direct communion of potential candidates with small groups of voters places enormous difficulties in the way of those party leaders disposed to work beyond the primaries to the general election and to put forward the most appealing slate. Individual politicians with a grasp on a small bloc of voters which can be turned into a primary victory are difficult to discipline or to bargain with. The support of even a weak personal organization, the loyalties and admiration of an ethnic group, a wide acquaintance within a religious group, simple notoriety achieved in a variety of ways, an alliance with an influential newspaper—these and a variety of other elements may create power within the narrow circle of people who share control in the politics of the direct primary.[15]

All these considerations bear on the capacity of the political parties to perform the function of nominating presidential candidates in a manner that reflects a consensus of some population that by ordinary criteria of democratic theory ought to have the right to nominate. Because the system that primaries largely replaced was also flawed in its fulfillment of precepts of democratic theory, observers and participants may well differ about which system overall is better. My purpose here is merely to indicate that the matter turns out to be a closer question than is commonly believed to be the case.

Thus, I do not conclude that primaries should be discarded because they are only illusorily democratic. Rather, I suggest that democratic theory presents a mixed verdict with respect to primaries as a device for revealing and executing the popular will, and that not all versions of democratic theory make primaries so desirable as to bar the examination of a system dominated by primaries from the standpoint of its practical consequences.

2. Other Criteria: Peer Review and Deliberation

Such great stress has been put on the desirability of increased participation in the presidential nomination process that other possible desiderata have been neglected. It is well, however, to consider a few other possibilities. The design of a presidential selection process might, after all, in the abstract reflect a number of values associated with democratic government. From time to time a wide variety of practical mechanisms have been employed or proposed, and all have weaknesses. For example:

(1) Selection of the nominee by the party's congressional caucus. This process neglects the interests of those parts of the country in which the party is weak, and which fail to return members of Congress for the party in question. It biases the choice toward candidates known by, favorably known by, and perhaps especially accessible to members of Congress, themselves an elite group focused to a far greater degree today on Washington and its politics than was the case when the congressional caucus actually selected the nominee.

(2) Selection by a national primary. This process would pre-

sumably favor whatever candidates were well known to citizens at large and who could, for whatever reasons, mobilize a strong factional following. In a large field of
candidates, as a national primary would surely attract, the
two surviving to the run-off would presumably be the two
with the most tenacious hold on the largest factions, but
this might produce a set of alternatives deeply disapproved
of by many—conceivably even a majority—of the party's
voters.

In some respects, these two mechanisms represent polar alternatives, whose weaknesses are complementary. The caucus
fails to represent enough legitimate interests in the population,
and thus is too narrow; the primary, while in theory broadly
participatory, is in fact a lottery without the capacity to deliberate, which in consequence is liable to set at naught the participation of most party voters.

What criteria ought a presidential selection process to meet?
Two technical desiderata immediately suggest themselves:

(1) The process has to be able to meet deadlines imposed by
    the need to campaign before the general election takes
    place, and so it cannot be so elaborate as to come to no
    timely conclusion.
(2) And of course it must succeed in picking a candidate. It
    cannot result in a tie, or in an expression of indifference as
    among several alternatives.

To these, we might add two desiderata of a broad political
character:

(3) The process should do as little as possible to harm the sub-
    sequent chances of the party's candidate in the general
    election, or the party's capacity to field and elect candidates
    thereafter.
(4) It should select candidates who are likely to be able to exe-
    cute the duties and responsibilities of the office of Presi-
    dent with some exceptional degree of distinction.

Thus, simultaneously, the process should be in some sense good for the party, good for the system of competition between the parties, good for the Presidency, and good for the successful maintenance, more generally, of democratic government in the United States.

There seems to be little controversy about any of these as abstractly desirable criteria. Moreover, the system both as currently and as recently constituted does fulfill the technical criteria without great difficulty—though not without complaints associated with the allegedly too-early start of the election season. With respect to the next two criteria a number of practical problems have arisen owing to disagreement over proper instruments.

Consider the proper role of popular participation in the nomination process. In 1968 it could have been argued—indeed it was argued, most strenuously—that bars to full and-timely participation by rank and file members of the Democratic party caused disaffection that drastically harmed the party's capacity to contest the presidential election. Yet, after the party passed through a series of reforms, greatly enhancing participation in the selections of 1972 through 1980 the Democratic party was, evidently, even more incapacitated than before from contesting the general election.

Thus enhanced participation is no longer as popular as it once was as the sole criterion by which a party nomination process can be evaluated. At least two other criteria have been added to the list: peer review and deliberation. In some accounts these criteria are interchanged, but for our purposes it may be useful to distinguish them.

Peer review is a criterion which entails the mobilization within the party of a capacity to assess the qualities of candidates for public office according to such dimensions as intelligence, sobriety of judgment, intellectual flexibility, ability to work well with others, willingness to learn from experience, detailed personal knowledge of government, and other personal characteristics which can best be revealed through personal acquaintance. Whereas the participation criterion contemplates increasing the legitimacy of the nominee within the party by involving large numbers of people in the choice process, the criterion of peer review seeks to increase the confidence of voters in the capacities of the nominee actually to execute the office of the Presidency effectively. Since the sort of information involved in peer review is never likely to be spread widely in the population at large, the building of this criterion into the process of candidate selection presumably entails an enlarged role for experienced politicians, who have a long-run stake in the party and who may also have the means of informing themselves in detail about the abilities in non-public situations of candidates for public office.

This information, it is rightly pointed out, was fully available to the political leaders who entered a smoke-filled room to pick Warren G. Harding. Politicians have been known actually to prefer lesser to better candidates, to act from motives other than the purely disinterested, and to be grossly mistaken in their judgments. All these considerations argue the wisdom of restraining the criterion of review by fallible and self-interested peers from becoming the sole method controlling the selection of candidates. They do not, however, rebut the proposition that there are some things that Presidents must do that people exposed to candidates only through the intermediation of the news media are unable to inform themselves about. And so while mistakes and misapplications are bound to occur in peer review as with any process, the alternative of providing no means for the assessment of potential Presidents by peers seems less desirable. Without the potentiality—or perhaps the threat—of peer review the incentives for prospective candidates to maximize name recognition and minimize public service—to cultivate the show horse rather than the work horse style—would become overwhelming. This would be true for the idealistic and the cynical alike. Connoisseurs of the process by which the dominance of the criterion of participation takes its toll on the public service of candidates can, if they like, watch the way in which absenteeism on key legislative roll calls among Senators running for President and members of Congress running for the Senate has spread to earlier and earlier periods in the run-up to elections.

If having the good opinion of colleagues and others intimately connected with government and politics means little or nothing to a candidate's chances for advancement, the nomination process then works at cross-purposes with the process of governing, which relies so heavily on accountability among elites. This may lead, in the first place, to inferior government, as persons unable to pass muster with their peers occasionally prove to be popularly attractive. In the second place, it may contribute to popular disaffection with government, as complaints about ineffective on-job performance filter down from Washington and interest group elites into the constituencies.

Deliberation as a characteristic of a presidential selection process should be considered separately from peer review, because it is not the peers of candidates alone who should, or who do deliberate. I take this criterion by which a choice process is evaluated to mean, simply, that the devices for aggregating individual preferences into collective choices maintain some capacity for receiving, manipulating and responding to a wider range of information than the first choices of participants.

The opposite of deliberation in this conception is, I suppose, automation, and the outcome that deliberation is supposed to ward off is a situation in which the formal properties of the choice process—the sequential order in which decisions are scheduled, the particular set of alternatives available to a given set of choosers, the number of decisions to be made on a given day—impose a result. It is no secret that the processes through which human interaction and communication are facilitated and impeded have a powerful role to play in determining the outcomes that are permitted to emerge. The criterion of deliberation simply highlights this fact and rates as more satisfactory procedures in which choosers can explore their own preference structures and those of other relevant actors, can avoid out-

comes weakly preferred by small majorities and strongly disliked by large minorities, can juxtapose a number of different
sets of alternatives before settling on a collective choice.

While maximizing deliberation may sound like a plea for the
resuscitation of meaningful national party conventions, practical observers of political trends will caution that conventions
may come so late in the nomination process as to be beyond
redemption. Communication by means of mass media coverage
of early events in the nomination process is now so efficient
and so swift that it is simply impractical for party delegates to
wait until they assemble face to face in a grand conclave to dis-
cover what they are thinking. This may well be true; consequently the criterion of deliberativeness, if it is to be applied at
all, must be applied to the process as a whole. We must ask to
what extent, and by what means, sensitivity to the varied concerns of legitimate participants in the choice process can be
maintained and closure on a nominee having widespread support achieved.

The need to devise a choice process which gives weight to
more than one value almost certainly requires some diversity
or mixture of mechanisms. It is probably too simple to associate
primaries exclusively with participation, party leaders and
leader-dominated caucuses with peer review and interaction
between or mixtures among delegates chosen by different
mechanisms with deliberation. As I have attempted to show,
primaries are imperfect mechanisms for participation; as the
notorious smoke-filled room suggests, peer review is imperfectly
accomplished by party bosses; and serious presidential selection
by face to face interaction among large numbers of party leaders at national conventions seems an irretrievably nostalgic
proposition.

It seems worthwhile, nevertheless, to examine a few proposals that have bubbled to the surface of debate about party reform in order to gain some sense of the range of practical possibilities for achieving varying approximations of success in
meeting criteria like the three we have explored.

3. Some Practical Proposals

Over the last decade a large number of practical proposals to improve the presidential nomination and election process have been given some attention. Some old chestnuts, like the abolition of the electoral college, have not been much discussed in connection with the nomination process, although by the logic expounded in Chapter Two it is at least arguable that abolition would draw more candidates into the election sweepstakes, and fewer of them would accept national party nomination as definitive. Pressure would build to rectify the financial advantages now legally conferred on the major parties in favor of a larger, ever more serious cadre of independent candidacies, and the capacity of the parties to channel preferences would, no doubt, suffer further reverses.

Within the parties themselves, conversation has mostly focused upon strengthening rather than weakening the influence of party in the nomination process. Many suggestions have been made, at least six of which seem worth discussing.

A. *Windows.* An unexpectedly vehement body of opinion has developed among party leaders—including leaders in the vanguard of the last round of reform—on the subject of primary elections.[16] Complaints have largely been made about excessive expenditures of money, time, and energy, about a process that stretches out too long, about how the early primaries totally dominate the process and obliterate the significance of later ones in spite of the incapacity of any small cluster of states to "represent" all the rest, never mind tiny New Hampshire with its elderly population, its large French-speaking minority, its backward economy, and its lack of state-wide communication other than the wildly idiosyncratic *Manchester Union-Leader.*

Yet the morning after the New Hampshire primary, politicians complain, something snaps into place on the NBC Today show that no amount of hard work in Massachusetts or New York or California can seem to rebut.[17] The proposal to force delegate selection—by primary or otherwise—into a shorter time period, a "window" of perhaps three to five months, is an attempt to meet this problem.[18] If three, or seven, or twelve states all selected their delegates on the same early day, it is supposed that the resulting challenge to the television networks' capacities for sorting signal from noise would keep a few more candidacies alive a little longer. The state-by-state party processes of delegate selection could be allowed a little more leeway to proceed unmolested by media-induced bandwagons and a little power would have been snatched away from the news media and vested in the state parties.

On the other hand, it might not work that way, and the overwhelming need for reportorial coherence might simply mean that the first trial heat, consisting of several states, and hence a more legitimate test than New Hampshire alone, would totally overshadow the rest. Soon, all states would clamor for early entry[19] and by swift increments a national primary would be the de facto result with its overwhelming reliance on participation to the exclusion of all other values.

If the proposal works as its proponents hope, there are still problems. Lone Ranger challenges such as the one Senator Eugene McCarthy launched in 1968 would be far more difficult if it had to be expensively conducted in several, conceivably widely scattered, states all at the same time. So the real losers in such a scheme might be long-shot candidates, and the real winners the pre-primary favorites. Insofar as pre-primary favoritism can still be conferred by the good opinion of the mass media, their powers would be unimpaired. Practically speaking, however, the news media, before the various straw polls and primaries on which they peg their coverage are held, tend to "take seriously" candidates who are well regarded by virtue of peer reputations among heavyweight national politicians. So the adoption of the window plan may enhance peer review in the process, would hurt the prospects of an effective protest being mounted through the primaries, and therefore would likely as not aid mainstream and centrist candidacies.

B. *Reserved seats at the convention.* A method for reintroducing peer review, and perhaps deliberation, into the national party convention might be to reserve delegate seats for various public or party officials.[20] In the most common version of this proposal these officials would not be required to commit themselves to some candidate in advance, as virtually all other delegates now generally do. If the proportion of uncommitted to committed delegates is high enough, presumably these uncommitted officials would be in a position to take the lead in turning the convention into a genuinely deliberative body, in which the criterion of peer review also would be given due regard. Much, therefore, depends upon the ratio of uncommitted to committed delegates. It is difficult to specify a threshold number in advance of some accumulated experience. Below the threshold, ex-officio delegates would almost certainly refuse to hold out against the favored candidate of the bulk of committed delegates. Public and party officials must, after all, work with the presidential candidate and certainly with the president of their party. Moreover, the elected officials among them have their own election and reelection to worry about and are unlikely to be interested in quixotic gestures. It seems improbable that either party will soon try to send to the national convention a contingent of uncommitted elected and party officials large enough to provide cover for a deliberative process at the convention that can afford to be seen to thwart the desires of delegates selected in the ordinary way. For these reasons, what appears at first blush to be a means for reasserting the values of peer review and deliberation promises in practice to be ineffective.

The definition of democracy as increased participation measured simply by numbers of participants is straightforward and appealing to many commentators in the mass media. Moreover it has the incidental effect of increasing the influence in the process of the mass media, which are best at reaching large numbers of people regardless of their affiliations or commitments to party organizations. Thus a political analysis of the process of party reform, and of recurrent episodes of reconsideration of reform, would suggest that insofar as the conceptions of democracy and fair play most commonly entertained by news media commentators are invoked, they come in on the side of the post-1968 wave of changes, which "opened up" the process, "democratized" it, and provided for greater participation.[21]

Within the political parties, the fundamental division of interests is between presidential candidates and their agents on the one hand, and state parties and their leaders on the other. It is in the interests of prospective presidential candidates for state delegations to be relatively permeable to their influence, for decisions about the identities of delegates to be made in highly visible processes, for commitments to candidates to come early in the selection process. It is in the interests of state parties and their leaders to maintain their influence over the composition of delegations to the national conventions, to remain neutral as among possible candidates until they can see which way the wind is blowing and arrange for appropriate access for the benefit of the state party.

This division of interests seldom results in a clear-cut debate over the legitimacy of the two sides: whether prospective presidential candidates and their minions ought to dominate party reform commissions, whether state party leaders ought to be the people represented on party decision-making bodies. Good politicians in the American tradition do not seek confrontations on issues of this kind. Nevertheless, this fundamental issue—the issue of who the President-selecting party is and ought to be—pervades the politics of party reform and needs clarification.

Opinion within the Republican party is far less divided than among the Democrats on this issue, principally because the most popular Republican presidential candidates and the most influential state party leaders have typically not been far apart on political issues during the 1960s and 70s. Among the Democrats, however, the rejection of state party leaders turned out to be the structurally significant consequence of the attack from within the party on Lyndon Johnson and his war policies. It has proven difficult to dislodge presidential candidates from their position of influence and to reassert the legitimacy of state party leaders. This, certainly, is one way to read the pulling and hauling that has taken place in the several Democratic party re-reform commissions—Mikulski, Winograd, Hunt—over the entitlement of state party leaders and elected politicians to get a free ride as uncommitted delegates to national party conventions. Uncommitted delegates are scarcely what prospective presidential candidates want, and so the candidates and their agents on reform commissions battle to minimize their numbers. Until the numbers of delegates uncommitted for one reason or another are large enough to return to the national party convention a genuine decision-making function in the presidential selection process, the submergence of state parties and their leaders, and the dominance of presidential candidates and their agents will continue.

C. Binding or unbinding delegates. Party rulemakers have tinkered with various formulas for requiring delegates to vote for the candidate they agreed to support during the process that selected them, or, contrariwise, not requiring that delegates do so. Much of this discussion, as I have mentioned, has been difficult to follow owing to the general failure to sort out the underlying question of why various delegates are present at the convention in the first place. In most states "delegates for" various candidates are apportioned to candidates on the basis of the candidate's proportion of the primary vote. The delegates themselves are sometimes picked by candidate managers after it is determined how many bodies are needed, and then these slots are filled; sometimes they are picked beforehand, and run "committed" to candidates of their choice. In the case of the Democrats, at large and post-primary delegate vacancies are filled according to the complex rules of demographic identity that are now written into party law. By such straightforward means black delegates for George Wallace can be sent to national conventions.

Insofar as delegates come into convention as designated representatives of the candidates who selected them, whether the convention itself acts to bind them or omits to do so seems immaterial. Insofar as delegates represent various interest groups in some sort of alliance with candidates it may be to the candidates' advantage—though surely not much to their advantage—to bind them by convention rule. If delegates were, as they increasingly are not, authentic representatives of state party organizations, then of course the flexibility of a lack of commitment might be of benefit to them in case no candidate came into the convention with a majority of committed delegates. But even if a delegation consisted of nothing but state party loyalists, it is hard to see how it could claim the right to maneuver freely if it had made a commitment to a candidate as a condition of its selection by a state party caucus or primary.

The issue, in short, cannot be settled at the convention but only at the state level where the original bargain between candidates and delegates is sealed. This is the case, at least, in theory. The legal niceties on this issue are not materially clarified by what seems to be a current Supreme Court view that whatever takes place at a national convention is subject in the first instance to national party rules rather than state law.[22] As a political matter the binding of delegates seems almost an irrelevancy given the methods by which they are most commonly chosen.

D. *Increasing allowable financial contributions.* As inflation proceeds apace, the limitations to which givers of political money are restricted means that contributions can buy less and less. Independently collected and expended funds which escape regulation take on greater importance. The capacity of parties to conduct their own business declines. Thus a general consensus has developed favoring the setting of contribution limitations at some higher level. The advantages of dispensing altogether with these limitations, as well as limitations on expenditures, is less often discussed, even though the case for no limits at all seems uncomplicated and compelling.

Public finance of elections serves two clear purposes. It encourages political competition by helping serious major party candidates get into the race, and it provides a framework for the disclosure of receipts and expenditures of money in political campaigns. Limitations on contributions and expenditures are, at best, attempts to equalize the financial resources available in primaries to various candidates. They have failed to do so. Rich candidates can and do outspend poor ones because so much of the personal funds of candidates are beyond the reach of regulation.[23] And well-publicized candidates can generally raise more money in small contributions than badly publicized ones. Independent expenditures cannot be regulated, but since candidates and parties can be, the influence of independent political operators increases at the expense of parties and candidates. Monitoring of expenditures and contributions creates great amounts of regulatory business and requires defensive measures by candidates which detracts from their central tasks of campaigning.

The original justifications for limitations consisted mostly of gossip and hearsay about the alleged susceptibility of candidates for the Presidency or electorates to be bought.[24] The persistence of these beliefs makes it highly unlikely that expenditure or contribution limitations will be abolished. More likely they will be adjusted upward, but the structure of regulations will remain and the power of organizations outside the reach of regulation will continue to grow.

E. *Abolition of cross-over primaries.* In a few states, it is possible for voters to vote in the primary of their choice without making a prior declaration of party membership. A decision of the Supreme Court places firmly in the hands of the national party the right to exclude delegates chosen by an electorate contaminated in this fashion.[25] No doubt national parties will be tempted to exercise this right. The effects on any given presidential race will be utterly trivial. National parties can, however, with impunity inflict needless embarrassment upon the state party of a state such as Wisconsin, which maintains a cross-over primary as a survival of its progressive political heritage. The national party that hastens first to do this—it will probably be the Democrats, who joyfully did battle with the state of Wisconsin in the Supreme Court—undoubtedly will suffer marginal losses in popularity in the state.[26] This sort of reform provides a pristine example of an excellent principle pointlessly—indeed fatuously—pursued. It gives a bad name to reform.

F. *Primaries detached from delegate selection.* A final suggestion dramatically cuts the Gordian knot that the primaries have tied around the nomination process and proposes that states be permitted to select their delegates by whatever means they choose, but that they be required to run an advisory or beauty contest primary. Thus the voters of each state would have a chance to express their views to their respective state delegations, but state delegations would retain the right to maneuver as they pleased. This proposal leaves open the issue of delegate commitments to candidates, though presumably it does not favor them. The actual prevention of prior commitments could be effected only if the delegate selection process were insulated in ways that McGovern guidelines make difficult. This proposal also provides for a restriction on the scope of participation while increasing the amount of voting. It seems unlikely to be enacted, in the first place because many states have primary elections that actually select delegates as a legitimate part of their customary political practices. The main losers if this proposal were enacted would undoubtedly be the candidates, since they would have to sweep advisory primaries in a fashion approximated in 1952 by Dwight Eisenhower in Minnesota in order to carve out something like the sort of autonomy from state party leaders that they presently enjoy. It is, of course, an attenuation of that autonomy that is at the heart of this proposal and that commends it to its advocates.

4. On Turning Back the Clock: Are Parties Recoverable?

Nobody knows with certainty whether any amount of tinkering with the presidential nomination process will in the end do the slightest good. The most common responses to attempts to restart the engine of party in presidential nominations are assertions about the irreversibility of events. In such discussions metaphors about the futility of turning back the clock, returning the genie to the bottle and toothpaste to the tube abound.

It therefore seems desirable to explore the assumption, implicit in much of this book, that political parties are worth the trouble we have been taking over them, that they are institutions whose functions are so vital as to make their preservation and vigor a matter of legitimate concern. It is certainly the case that much of what political parties have accomplished in the past is now done, on the whole with greater evenhandedness, efficiency, and dignity by other means. Welfare systems, for example, see to the indigent and disadvantaged without imposing quid pro quo restraints on the political attitudes or voting behavior of recipients. Civil service systems at the municipal, state, and federal levels show a capacity for recruiting job holders according to expected quality of job performance to a degree never contemplated in most patronage systems. Schemes for competitive bidding on service contracts save taxpayers large amounts of money over the sweetheart deals that Plunkitt of Tammany Hall once described as "honest graft."[27]

So, undeniably, the range of activities that political parties once undertook to perform has shrunk steadily. This general shrinkage of the scope of party activity is not a phenomenon one associates with the transformations of the post-1968 era. The professionalization of public service has been going on at least since the assassination of President Garfield in 1881. Plunkitt's phillipics against the civil service were uttered at the turn of the century.[28] Modern institutions of welfare are on the whole in the United States a much more recent phenomenon, but it is in any case unlikely that even the most robust partisan of strong local parties would wish the return of the indigent populace to dependency upon gifts of food from their precinct captains. Nostalgia, in short, will not do as a foundation for public policy, and especially since there can be so little grounds for regret in the changes that have taken place.

On what basis, then, can political parties be recommended
for a continuing role in the nomination process? Perhaps two:
no better institutions have evolved to conduct nominating politics, and, second, the tasks parties still perform are both essential to harmonize with the presidential nomination process and
crucial for the proper general functioning of the political system. Parties are organizations that, when they are functioning,
mobilize voters, coordinate the activities of leaders, and by recruiting candidates and sponsoring political campaigns, provide some linkage between the two levels of politics. These typical intermediary activities can in part be performed by interest
groups and by the mass media, but none of these tasks is central to the lives of interest groups and to the mass media as
organizations. If either type of organization actually found itself stuck with the ongoing responsibility of performing classic
party functions it would soon either disintegrate, reshape itself,
or grow ancillary structures. Demands for access and accountability from unfamiliar sources would have to be met and dealt
with, and in many cases acceded to, to avert electoral defeat.
For, exclusive, narrow, and hierarchical as various state and local
parties may be, seen from the perspective of the nation as a
whole, American political parties are broadly based and highly
diversified structures when compared with any interest group
or cluster of interest groups. They are able to generate policy
on a very wide range of topics, and to provide a legitimate
forum for stalemate on topics where there is disagreement. The
capacity of parties to gather and process information about human wants and demands nation-wide is impressive even when
compared with the wire services and the television networks,
and their capacity to evaluate wants and demands and select
among them in some coherent and organizationally defensible
fashion greatly exceeds that of any of the news media.

Given the formidable needs for coordinated political activity
required in the face of the constitutional separation of powers,
of the federal structure of the nation; of its geographic reach,
of the heterogeneity of section, race, ethnicity, occupation, and

economic condition that characterizes the American population, it is fair to say that parties still have work to do and were they not to exist, something very like them would have to be invented. Consequently recommendations concerning the organizational design and worries about the organizational performance of parties have some considerable point. It is by no means clear that the revival of organizational integrity among political parties entails turning back the clock to conditions more hospitable to their organizational survival; potentially hospitable conditions exist today and seem likely to continue to exist so long as leaders continue to be elected by popular vote and legislatures meet in the fifty states and in Washington. Presidents, governors, and mayors will continue to have the need for voters who elect them and the legislators on whom they must rely organized in some fashion. Political party seems as good a name as any to give to the organizations that undertake these tasks.

Thus what is at issue is not the survival of basic party functions so much as the thoughtful design of organizations to perform those functions in as effective a fashion as possible. What seems to have happened, looked at in a very broad overview is simply that an organization that works best when it has been highly decentralized has been significantly centralized in its powers and operations.

Historically, this centralization process can be interpreted as the last poisonous gift of the American institution of slavery. Slavery discredited all the decentralized organizations that fed on it and protected it and its successors. As late as 1964 a transparently racist state delegation from Mississippi was sent to the Democratic national convention, representatives of the dying politics of intimidation and denial of the right to vote on racial grounds.[29] Because the rest of the Democratic party for all sorts of reasons could not countenance that, justifications had to be found and systematically argued on general principles to the effect that the national convention had the power to keep its own house in order and to enforce rules of decency upon its component parts, the state parties. The same general principles, asserting the simple right of the whole to regulate the part, can serve equally well to forbid the innocuous Wisconsin cross-over primary.

There is nothing incoherent in constitutional doctrines which proclaim an institution capable of asserting centralized standards in one matter, and likewise in another. So further appeal in the Wisconsin case, if one should be forthcoming, cannot stand on constitutional grounds, but must on prudential grounds. As a matter of prudence, not jurisprudence, the national conventions, their rules committees, and the reform and re-reform commissions which seem so readily to be adopted by party bureaucrats as creatures of the conventions, should be reluctant to impose any standards at all on the parties of the several states. Racism, that ugly legacy of slavery, should be the great and obvious exception. Where racism survives in the conduct of local party business the representatives of all the other state parties should on prudential grounds exercise the power the courts have given to them to take whatever measures are needed to assure fair, equitable, and race-blind political participation. On prudential grounds all other interventions into state and local politics, all other attempts to make uniform the political cultures of widely separated, historically distinct entities, ought to be resisted. The prudential grounds are these: state and local parties that root in local soil have the best chance of surviving and prospering, of winning elections and recruiting good candidates, and bringing to the national nominating process the authentic voices of the people of the several states. These voices deserve to be heard, and it cannot be assumed that the formulas that work in eliciting these voices in a place like California also will work in the same way in Louisiana or Vermont or Ohio.

If central conceptions of justice and equity should not entirely determine the contributions that state parties make to presidential nominations, what should? Even in places where local patronage organizations still flourish, it does not seem unreasonable to conclude that representation at national party conventions should follow the social contours that influence the ways in which local party effort is organized. States that are nonpartisan in spirit will rely more heavily on middle-class volunteer workers, and this spirit ought freely to be permitted to emerge at national party conventions as well. Interest groups at the local and state level ought to be given incentives for involving themselves in alliances with local and state parties. State and local parties can be greatly strengthened in their state and local political efforts by such alliances, and these alliances can be further facilitated by the assurance that state and local parties—not deals directly with presidential candidates—will form the basis for participation in the presidential nominating process.

In some states, to be sure, where state and local parties are weak, such deals with national candidates will continue to be made. The strongest overall design, the design that best helps nominees to get elected and Presidents to govern, is that of the mixed system: state conventions dominated by party bosses where party bosses survive and actually coordinate local political effort; interest group dominance of presidential nominations where interest groups finance and staff campaigns; and primary elections where parties are weak or where the norms of civic participation so dictate.

Do mixed systems necessarily produce better candidates? In most respects the question is unanswerable, but not, perhaps in one. Mixed systems by expressing the varied political cultures of the far-flung United States provide a better education to prospective candidates than any more uniform alternative choice process. Candidates who must come to terms with the realities of American politics in all their variety and complexity are better suited to governing America than those for whom such an education is unavailable or unattainable. Mixed systems may pick a different winner than more uniform processes in a given presidential year or they may not. What they can be assured of doing year in and year out is placing within each candidate's grasp a more enlightened understanding of what it takes to govern.

Notes

Preface

1. For a list of some of them see Nelson W. Polsby, "Contemporary Transformations of American Politics: Thoughts on the Research Agendas of Political Scientists," *Political Science Quarterly* 96 (Winter 1981–82), pp. 551–70.
2. *Presidential Elections: Strategies of American Electoral Politics* (New York: Scribner, 1964, 1968, 1971, 1976, 1980).
3. Byron E. Shafer, "The Party Reformed," Ph.D. dissertation, University of California, Berkeley, 1979; to be published as *The Quiet Revolution: Reform Politics in the Democratic Party, 1968–1972* (New York: Russell Sage Foundation, forthcoming).
4. "Presidential Cabinet Making: Lessons for the Political System" (Bloomington, Indiana: The Poynter Center, June 1977) and *Political Science Quarterly* 93 (Spring 1978), pp. 15–25; "Coalition and Faction in American Politics: An Institutional View," in S. M. Lipset (ed.), *Emerging Coalitions in American Politics* (San Francisco: Institute for Contemporary Studies, 1978), pp. 103–23, and Lipset (ed.), *Party Coalitions in the 1980s* (San Francisco: Institute for Contemporary Studies, 1981), pp. 153–78; "Interest Groups and the Presidency: Trends in Political Intermediation in America," in Walter Dean Burnham and Martha Wagner Weinberg (eds.), *American Politics and Public Policy* (Cambridge, Mass.: MIT Press, 1978), pp. 41–52; "The News Media as an Alternative to Party in the Presidential Selection Process," in Robert A. Goldwin (ed.), *Political Parties in the Eighties* (Washington, D.C.: American Enter-

prise Institute, 1980), pp. 50–66; and "Party Reform and the Conduct of the Presidency," in James Sterling Young (ed.), *Problems
and Prospects for Presidential Leadership in the 1980's* (Lanham,
Maryland: University Press of America, forthcoming).

Introduction

1. In 1980, for example, that bellwether of American liberal journalism, Tom Wicker of the *New York Times,* said of the two candidates of the major parties:

   Think of that: an incumbent President whose record of ineptitude stands
   unmatched since Warren G. Harding and whose campaign is based on
   foreign policy crises largely of his own making; or a 69-year-old ex-Gov-
   ernor twice rejected by his own party, with no foreign policy, national
   security or Congressional experience.

   "Getting Down to Cases," *New York Times* (March 16, 1980). As is
   frequently the case, Mr. Wicker was not alone. A sampling of
   complaints reaching wide audiences might include: Albert R. Hunt,
   "The Process Is Out of Kilter," *Wall Street Journal* (November 10,
   1980); Cyrus R. Vance, "Reforming the Electoral Reforms," *New
   York Times Magazine* (February 22, 1981), pp. 16ff; Adam Clymer,
   "Poll Finds Reagan-Carter Choice Unsatisfactory to Half of Public," *New York Times* (April 18, 1980); Donald C. Bacon, "What's
   Wrong with the Way Americans Pick a President?," *U. S. News and
   World Report* (March 31, 1980), p. 35; Richard Reeves, "When Reform Backfires," *Esquire* (March 1980), pp. 8, 11; "Is This Any
   Way To Pick a President?," *Newsweek* (October 15, 1979), pp. 69–
   70, 79; and Terry Sanford, *A Danger of Democracy: The Presidential
   Nominating Process* (Boulder, Colo.: Westview Press, 1981).
2. Carter and Reagan's only competitors are Grover Cleveland
   (elected in 1884) and Woodrow Wilson (elected in 1912), neither
   of whom served in national government before becoming President. All thirty-five other Presidents had at least some prior national service to their credit, and for most that service was extensive.
3. Even so, by mass criteria both major party candidates were uniquely
   unpopular in 1980. For the first time, a University of Michigan
   survey recorded that both major party candidates were on balance
   viewed unfavorably by the mass electorate. See Arthur H. Miller
   and Martin P. Wattenberg, "Policy and Performance: Voting in
   the 1980 Election," delivered at the 1981 Annual Meeting of the
   American Political Science Association (pp. 5–6, figure 1).

4. See Alan L. Otten, "Western Europe Is Less Than Thrilled by Carter-Reagan Choice in U.S. Vote," *Wall Street Journal* (June 25, 1980). Among the European expressions of what Otten characterized as "near horror" about the 1980 presidential choice were these: "*Le Monde* asks: 'How can the most powerful democracy in the world find itself reduced to a choice between candidates of so little weight?' " and, from a British journalist, " 'We give our actors knighthoods, but we don't make them prime minister.' " Anthony King says: "All over Europe in the autumn of 1980, wherever people met to talk politics, there was only one topic of conversation: How on earth had a great country like the United States, filled with talented men and women, managed to land itself with two such second- (or was it third- ?) rate presidential candidates as Jimmy Carter and Ronald Reagan?" "How Not To Select Presidential Candadates: A View from Europe," in Austin Ranney (ed.), *The American Elections of 1980* (Washington, D.C.: American Enterprise Institute, 1981), pp. 303–28.

5. How presidential campaigns are financed I take to be a fundamental issue of party finance, even though, strictly speaking, it is candidates for public office, not party organizations per se, that are raising and spending most of the money with which we shall be concerned.

Chapter I: The Party Reforms and How They Grew

1. A thorough account of the presidential selection process in 1952 is contained in Paul T. David, Malcolm C. Moos, and Ralph M. Goldman, *Presidential Nominating Politics in 1952,* 5 volumes (Baltimore: Johns Hopkins University Press, 1954). See especially, Volume 1, *The National Story*.

2. See Mary Earhart Dillon, *Wendell Willkie* (Philadelphia: Lippincott, 1952), pp. 311–35.

3. See Jules Abels, *Out of the Jaws of Victory* (New York: Holt, 1959), pp. 57–61; and Irwin Ross, *The Loneliest Campaign* (New York: New American Library, 1968), pp. 47–53.

4. *The National Story,* pp. 31–33.

5. He won in California, Illinois, Maryland, Massachusetts, Nebraska, New Hampshire, New Jersey, Ohio, Oregon, Pennsylvania, South Dakota, and Wisconsin. He entered 14 out of the available 16 primaries, and won 12 of them (*The National Story,* pp. 55–56).

6. *Ibid.,* p. 65.

7. *Ibid.,* pp. 150–55.

8. President Truman discusses his views of the various candidates in *Years of Trial and Hope, 1946–1952* (Garden City, N.Y.: Doubleday, 1956), pp. 490–97.

9. A group of labor leaders representing both the AFL and CIO met with the 74-year-old Barkley at breakfast on the day before the convention and announced, with regret, that they could not support him because of his age (*The National Story,* p. 117).

10. This very complex business is described in *The National Story,* pp. 136–49; in John Bartlow Martin, *Adlai Stevenson of Illinois* (Garden City, N.Y.: Doubleday, 1976), pp. 593–95; and in Jack Arvey, as told to John Madigan, "The Reluctant Candidate," *The Reporter* (November 24, 1953), pp. 19–26.

11. See Martin, *Adlai Stevenson,* pp. 597–98; *The National Story,* pp. 151, 154.

12. Martin, *Adlai Stevenson,* pp. 513–604, is exhaustive on Stevenson's hesitations.

13. See V. O. Key, Jr., *Politics, Parties, and Pressure Groups,* 5th edition (New York: Crowell, 1964), pp. 375–94; Louise Overacker, *The Presidential Primary* (New York: Macmillan, 1926); Paul T. David, Ralph M. Goldman, and Richard C. Bain, *The Politics of National Party Conventions* (Washington, D.C.: Brookings Institution, 1960), p. 249.

14. Joseph Bruce Gorman, *Kefauver: A Political Biography* (New York: Oxford University Press, 1971), pp. 80–85.

15. Charles A. H. Thomson reports that an unpublished NBC survey of television viewers of the 1952 Democratic Convention in the New York area showed only Kefauver supporters responding that television was the most important news medium (ahead of newspapers) in helping them make up their minds. *Television and Presidential Politics* (Washington, D.C.: Brookings Institution, 1956), pp. 47–48. Other evidence of Kefauver's high and favorable visibility (this time among television viewers in southern Ohio) is found in, Members of the Department of Marketing, Miami University, Oxford, Ohio, *The Influence of Television in the Election of 1952* (Oxford, Ohio: Oxford Research Associates Inc., December 1954), p. 22.

16. Gorman, *Kefauver,* p. 88.

17. *Ibid.,* p. 91.

18. *Ibid.,* p. 91. This comparison is all the more remarkable in view of the fact that a local team, the Yankees, played in this series, beating the Philadelphia Phillies four games to none.

19. *Ibid.*, p. 87.
20. *Ibid.*, p. 92.
21. *Ibid.*, p. 92. See also G. D. Wiebe, "Responses to the Televised Kefauver Hearings," *Public Opinion Quarterly* 16 (Summer 1952): "The routine life of [New York] city was substantially altered as people interrupted normal pursuits to sit and watch the parade of local corruption and bribery that was unfolded on their television screens" (p. 180).
22. Gorman, *Kefauver*, p. 106.
23. Republican leaders in 1952 would have been fortunate to have had available Herbert H. Hyman and Paul B. Sheatsley's, "The Political Appeal of President Eisenhower," *Public Opinion Quarterly* 19 (1955–56), pp. 26–39. Such evidence as the Minnesota primary proved to be equally effective for their purposes.
24. An attempt to codify textbook rules of the nomination process as of 1960 is Nelson W. Polsby, "Decision-making at the National Conventions," *Western Political Quarterly* 13 (September 1960), pp. 609–19, esp. 616.
25. Theodore H. White, *The Making of the President 1960* (New York: Atheneum, 1961), p. 87.
26. For accounts of the Smith-Hoover contest of 1928, see Ruth C. Silva, *Rum, Religion, and Votes: 1928 Re-examined* (University Park: Pennsylvania State University Press, 1962); and Allan J. Lichtman, *Prejudice and the Old Politics* (Chapel Hill: University of North Carolina Press, 1979).
27. White, *1960*, p. 85.
28. *Ibid.*, pp. 97–114; and Harry W. Ernst, *The Primary That Made a President: West Virginia, 1960* (New York: Holt, 1962).
29. See Ithiel de Sola Pool, Robert P. Abelson, and Samuel Popkin, *Candidates, Issues and Strategies* (Cambridge, Mass.: MIT Press, 1964), pp. 68, 117–18; Philip Converse, Angus Campbell, Warren E. Miller, and Donald Stokes, "Stability and Change in 1960: A Reinstating Election," *American Political Science Review* 55 (June 1961), pp. 269–80; and Philip Converse, "Religion and Politics in the 1960 Election" in Angus Campbell (ed.), *Elections in the Political Order* (New York: Wiley, 1967), pp. 96–124.
30. Aaron Wildavsky, "What Can I Do? Ohio Delegates View the Democratic Convention," in Paul Tillett (ed.), *Inside Politics: The National Conventions, 1960* (Dobbs Ferry, N.Y.: Oceana, 1962), pp. 112–31. DiSalle tells the story somewhat differently in his autobiography, *Second Choice* (New York: Hawthorn Books, 1966), pp. 197–202.

31. V. O. Key, Jr., *Politics, Parties, and Pressure Groups,* p. 403.
32. "Six-month U.S. Toll at 9557," *New York Times* (July 4, 1968); and Joseph B. Treaster, "Fresh Fighting Reported at DMZ: American Combat Deaths Pass the 30,000 Mark," *New York Times* (December 13, 1968). At the end of March 1968, according to President Johnson, 525,000 American military personnel were in Vietnam. See Lyndon B. Johnson, *The Vantage Point: Perspectives of the Presidency 1963–1969* (New York: Holt, 1971), p. 436.
33. *Statistical Abstract of the U.S. 1973* (Washington, D.C.: U.S. Government Printing Office, 1973), p. 148.
34. See Theodore H. White, *The Making of the President 1968* (New York: Atheneum, 1969), pp. 259–63, 271.
35. See Lewis Chester, Godfrey Hodgson, and Bruce Page, *An American Melodrama* (New York: Viking, 1969), pp. 21–31.
36. See Gallup polls for January 21, 1968, February 18, 1968, March 10, 1968. Also John E. Mueller, *War, Presidents, and Public Opinion* (New York: John Wiley, 1973).
37. See David Halberstam, "The Man Who Ran Against Lyndon Johnson," *Harper's* (December 1968), pp. 47–66.
38. Among the people approached by Lowenstein who gave the matter some consideration were John Kenneth Galbraith, Senator Robert Kennedy, Senator George McGovern, and Representative Don Edwards. See Jules Witcover, *Eighty-five Days* (New York: G.P. Putnam's Sons, 1969), pp. 27–29; David Halberstam, *The Unfinished Odyssey of Robert Kennedy* (New York: Random House, 1968), pp. 3–19; Jeremy Larner, *Nobody Knows* (New York: Macmillan, 1970), pp. 17–22.
39. See Eugene McCarthy, *The Year of the People* (Garden City, N.Y.: Doubleday, 1969), pp. 89, 294–95. See also Arthur Herzog, *McCarthy for President* (New York: Viking Press, 1969), pp. 75–99.
40. A leading McCarthy campaigner wrote: "When the votes were tallied, I was somewhat surprised that a full 48 percent of the voters liked LBJ enough to write in his name. But since all the newscasters said that our 42 percent was a great victory for McCarthy, we were ecstatic" (Ben Stavis, *We Were the Campaign* (Boston: Beacon, 1969), p. 26).
41. Witcover, *Eighty-five Days,* pp. 15–44; Halberstam, *Unfinished Odyssey,* pp. 3–67.
42. Quoted in Halberstam, *Unfinished Odyssey,* p. 67. The illegitimacy of a Kennedy dynasty was by no means conceded by those favorable to Kennedy. See, for example, Theodore White's incredulity at the hostility Kennedy encountered as he "tried to enter on his inheritance" (*1968,* p. 151).

43. See Halberstam, *Unfinished Odyssey.*
44. Lyndon B. Johnson, *The Vantage Point,* p. 435. See also *Congressional Quarterly Weekly Report* (April 5, 1968), pp. 731–34.
45. Narrative accounts now available on 1968 are deficient in their coverage of Humphrey and his camp. See, however, Meg Greenfield, "Hubert Humphrey in 1968," *The Reporter* (June 13, 1968), pp. 19–24; Hubert H. Humphrey, *The Education of a Public Man* (Garden City, N.Y.: Doubleday, 1976). Notably unfriendly to Humphrey are the discussions in Arthur M. Schlesinger, Jr., *Robert Kennedy and His Times* (New York: Ballantine Books, 1978); and Chester, Hodgson, and Page, *An American Melodrama,* esp. pp. 142–55. And, mesmerized as always by the Kennedys, is Theodore White, *1968.*
46. Humphrey, *Education,* p. 368.
47. *Ibid.,* p. 371. See also *Newsweek* (June 3, 1968), pp. 28–29.
48. Contrast the attitude of the *New Republic* in 1964 and 1968. In 1964 the *New Republic* said:

> With Humphrey as Vice President, the Senate will have a presiding officer who knows every nerve and muscle of that body; the President will have a traveling emissary welcome abroad because he is interested in the problems of others and unafraid of change; and the President will have too a collaborator who shares his hope that this good society can become "a great society" ["Johnson-Humphrey," *New Republic* (August 8, 1964), p. 4].

In 1968:

> The editors of this journal are among those who have applauded Hubert Humphrey's energetic and good-humored championship of the underdog for over two decades. We do not gladly turn against him now. If the clock could be turned back, we would turn it back. It can't be done. The Vice President is trapped in a desperate situation for which he is only partly responsible. But he is trapped. He could break out of it, but only by following Lyndon Johnson, removing himself from the race and turning his party free. . . . The fires burning in the hamlets of Vietnam and in the ghettos of our cities illuminate a face of America we would rather not see. The Vice President's complicity is inescapable; that is his tragedy ["Calling All Delegates," *New Republic* (July 27, 1968), pp. 3–4].

In *Me and Ralph* (Washington, D.C.: New Republic, 1976), p. 1, David Sanford paints a portrait of the management of the *New Republic* as of 1968 that will probably not surprise its readers:

> I remember very well the snowy night [in 1968] when I delivered galley proofs of a *New Republic* lead editorial exhorting Eugene McCarthy to run for President, to the home of Gilbert Harrison, then the magazine's owner. When I knocked at the door I was greeted by three children, Mrs. Harrison, and a yellow Labrador retriever. The youngest son called out, "Daddy, your friend is here," and in a second or two I saw Gil Harrison, bounding up the stairs from the basement with a look of great expectation on his

face that was dashed when he saw me. "Oh, Dave, it's you," he said with a faint air of disdain. As I stood in Harrison's book-lined study, still in my galoshes, the friend he was really expecting arrived at the door. It was Gene McCarthy, there to read the proofs.

49. Gallup poll, April 28, 1968.
50. Gallup poll, May 15, 1968.
51. The Gallup Report News Release, June 2, 1968, "Humphrey Choice of Local Democratic Leaders."
52. *Newsweek* (June 3, 1968), p. 28.
53. "Robert Kennedy's Chances: What a Survey Shows," *U.S. News and World Report* (June 3, 1968), pp. 48–49.
54. Kennedy ended the year having won primaries in the District of Columbia, Indiana, Nebraska, California, and South Dakota, receiving 30.6 percent of all the votes cast in primaries in 1968. McCarthy got 38.7 percent of all the votes cast in primaries, winning in Wisconsin, Pennsylvania, Massachusetts, Oregon, New Jersey, and Illinois. The remainder of the votes cast in primaries was widely scattered: 8.9 percent for unpledged delegates, 7.3 percent for Senator Stephen Young of Ohio, 5.1 percent for President Johnson, 3.2 percent for Governor Roger Branigan of Indiana, and so on. A convenient source is the Congressional Quarterly publication *Presidential Elections Since 1789* (Washington, D.C., 1975).
55. Richard Tuck, "Teddy We Hardly Know Ye," *Washington Post* (November 11, 1979).
56. See Jack Newfield, *Robert Kennedy: A Memoir* (New York: Dutton, 1969); Halberstam, *Unfinished Odyssey;* Schlesinger, *Robert Kennedy and His Times.* Also Richard Harwood, "The Robert F. Kennedy Legacy," *Washington Post* (June 6, 1969); Ronald Steel, *Walter Lippmann and the American Century* (Boston: Little, Brown, 1980), p. 587; Jules Witcover, *Eighty-five Days*, pp. 335–37.
57. Witcover, *Eighty-five Days*, pp. 322–23.
58. Humphrey, *Education of a Public Man*, p. 373.
59. See Stephen MacDonald, "Hovering Doves: If Humphrey Is Picked, Peace Movement Could Get Violent," *Wall Street Journal* (August 28, 1968); Meg Greenfield, "The Hyperbole of the McCarthy Supporters," *Washington Post* (July 28, 1968).
60. The Hughes Commission grew out of a meeting of the Connecticut State "McCarthy for President" Steering Committee, held in West Hartford on June 23, 1968. The participants in the meeting were unhappy about the outcome of the previous day's Connecticut state Democratic convention, where McCarthy supporters had

taken only 9 of the state's 44 delegate positions, even though the Connecticut delegation had been selected legally, and the number of McCarthy delegates was proportional to the number of McCarthy's supporters (200 out of 960) at the convention. The McCarthy group decided to assemble a formally neutral commission that could document their claims that, even if selection processes across the country were legal, they were neither fair nor democratic.

It was decided that for the commission to have its desired effect it would have to appear neutral and representative. After a great number of problems in securing members, the commission was constituted. It was composed of a chairman who was a McCarthy backer amenable to party reform (Gov. Harold Hughes of Iowa), a vice-chairman from the Humphrey camp (Rep. Donald Fraser), a black leader (Georgia state senator Julian Bond), a woman (columnist Doris Fleeson Kimball), a Kennedy supporter (California attorney Fred Dutton), and two well-known writers on political and legal subjects, Professor Alexander Bickel of the Yale Law School, a Kennedy supporter, and Harry Ashmore, formerly the executive editor of the Little Rock *Arkansas Gazette,* who had become an officer of the Center for the Study of Democratic Institutions in Santa Barbara, California, a McCarthy supporter.

The staff consisted mostly of Connecticut McCarthy campaigners. For example, Geoffrey Cowan, a Connecticut McCarthy delegate coordinator, became the associate staff director. He arranged with William M. Johnson, II, a small publisher who supported McCarthy, to contribute $10,000 in exchange for rights to publish the reports and whatever else might be written.

Tom Alder was selected as the commission staff director since Cowan seemed too visibly political, given his position in the McCarthy campaign. Alder had been less involved in the Democratic race because he was trying to start a fourth party to the left of the Democrats. The rest of the staff was composed of five McCarthy supporters, a Kennedy backer, and six persons who were unattached.

The staff had less than six weeks to assemble a report. On August 4, the co-chairman of the Connecticut McCarthy campaign, Anne Wexler, and Hughes publicly announced the creation of the commission. The commission met only once, on August 13. All the commissioners but Ashmore attended. At this meeting, commissioners made their only collective contribution to the commission report, agreeing to most of its staff's proposals.

Only two of these proposals caused the commissioners to balk. One was the staff's plan to abolish winner-take-all primaries. Dutton, of California, a winner-take-all state, objected. Second, the commission refused to do away with all state party conventions, as the staff also recommended even for those which could be considered "participatory." Fraser saved the latter.

The report, written by the staff, remained critical of the winner-take-all primary and the participatory convention, although in the end it was not recommended that they be abolished. The report recommended the representation of minority opinion at all levels of decision making, by ending use of the unit rule and winner-take-all elections. Public participation in the selection process was to be promoted, preferably through proportional primaries.

See Byron E. Shafer, "The Party Reformed," Ph.D. dissertation, University of California, Berkeley, 1979, pp. 8–64; to be published as *The Quiet Revolution: Reform Politics in the Democratic Party, 1968–1972* (New York: Russell Sage Foundation, forthcoming); Richard G. Stearns, "The Presidential Nominating Process in the United States: The Constitution of the Democratic National Convention," Ph.D. dissertation, Oxford University, 1971, pp. 25–39; and Commission on the Democratic Selection of Presidential Nominees, *The Democratic Choice* (Washington, D.C., 1968).

61. *The Democratic Choice*, pp. 22, 31, 33.
62. See Alan L. Otten, "Still the Boss: Many Detect LBJ's Touch in Convention Planning," *Wall Street Journal* (August 20, 1968); Laurence Stern, "HHH Sought To Shift Site of Convention," *Washington Post* (December 3, 1968); White, *1968*, pp. 338–39.
63. See "C.B.S. Man Is Attacked on Floor of the Convention," *New York Times* (August 28, 1968); "Guards Halt Delegate Till He Discards Paper," *ibid.;* Tom Wicker, "Victor Gets 1,761: Vote Taken Amid Boos for Chicago Police Tactics in Street," *ibid.* (August 29, 1968); Wallace Turner, "New York Delegate Dragged from Hall by Police," *ibid.;* "Scorn Expressed in Papers Abroad: Police and Security Actions in Chicago Assailed," *ibid.*
64. See Norman C. Miller, "Celebrity Delegates at Democratic Parley Get No VIP Handling," *Wall Street Journal* (August 28, 1968); White, *1968*, pp. 284–85; Chester, Hodgson, and Page, *An American Melodrama*, pp. 581–91.
65. The foregoing quotations are from *Rights in Conflict*, a report submitted by Daniel Walker, director of the Chicago study team, to the National Commission on the Causes and Prevention of Violence (New York: Dutton, 1968), pp. 1, 3–4, 50.

66. "In the Jungle," *New York Review of Books* (September 26, 1968), pp. 11–13. At this time, the *New York Review* was widely regarded as the most influential publication in the United States among intellectuals, according to a study of the subject by Charles Kadushin, *The American Intellectual Elite* (Boston: Little, Brown, 1974). For other reports on the Democratic convention, see Tom Wicker, "America Was Radicalized," *New York Times Magazine* (August 24, 1969), pp. 27ff.; Norman Mailer, "Miami Beach and Chicago," *Harper's* (November 1968), pp. 41–130; Elia Kazan, "Political Passion Play: Act II," *New York* (September 23, 1968).

67. White, *1968*, p. 302. See also, Stewart Alsop, "Can Humphrey Win?," *Newsweek* (September 9, 1968); Richard A. Pride and Barbara Richards, "Denigration of Authority? Television News Coverage of the Student Movement," *The Journal of Politics* 36 (August 1974), pp. 637–60.

Literary critic Elizabeth Hardwick offered the following mellow retrospective about a month later for the readers of the *New York Review of Books:*

Wednesday night, during the siege of the Hilton, when the police mercilessly beat young men before the eyes of everyone, you could hear the timid but determined voices of "concerned" women calling out, "What are the charges against that young man?" Or, "Stop, please, Sir, you are killing him!" The mention of the instruments of law and order sent the police into a wild rage and for a moment they stopped beating demonstrators and turned to threaten the frightened suburbans. During the raid on the McCarthy Headquarters, a girl in tears asked, "What are the grounds?" The police answered, "Coffee grounds." With this lawlessness of the Law, misery fell from the sky. Suppose, you found yourself wondering, *they* should take over! "I have been a life-long Democrat," people kept whispering in bewilderment. Few had realized until Chicago how great a ruin Johnson and his war in Vietnam had brought down upon our country.

This article was decorated with a David Levine cartoon showing Humphrey in a policeman's helmet, holding a truncheon. Ms. Hardwick continues:

Hubert Humphrey is an altogether embarrassing figure, with his dyed black hair and glowing television makeup. He creates a sense of false energy—like an MC on an afternoon show. The present Democratic leadership appears to be divided between bullies and cowards and Humphrey asks us to take our chances on the coward. You will find me less dangerous, he seems to be trying to assure us.

The Vice President has many words and he uses them over and over. "I am the Captain of the team," he says. Many of the choice sentences of his acceptance speech had been the choice remarks of his appearance before the California delegation. (Peace and freedom do not come cheaply, my friends.) He brought forth Winston Churchill and St. Francis of Assisi—one strong and one humble—and topped the embarrassment of the first by the second. He is always frantically smiling; repose is a rapid fade to sentiment. In between, where feeling and person would lodge, there is simply nothing. He does not seem in touch. Empty smiles, a wound-up toy. Nothing in him inspires confidence. He cannot allow himself to be distracted by events. The entire convention appeared to intrude upon his smiles. Nothing has happened since the Thirties: that is his message, that is the real Humphrey, now, "Captain of the team."

Elizabeth Hardwick, "Chicago," *New York Review of Books* (September 26, 1968), pp. 5–7.

68. Described in Hubert H. Humphrey, *Education of a Public Man,* pp. 383–94. See also, Louis Harris, "Conventions: Nixon Gained, HHH Was Hurt," *Washington Post* (October 3, 1968). McCarthy delegate John Kenneth Galbraith reminisces:

. . . There are times when one must be completely adamant, and this was one. The war was the transcendent issue. Compromise would not only have cost us the confidence of our own supporters, it would have provided sanction for those saying the war must go on. A firm stand would require Humphrey to move toward our position during the campaign.

A Life in Our Times (New York: Ballantine, 1981), p. 503.
69. *Congressional Quarterly Weekly Report* (August 30, 1968), p. 2287.
70. The chairman, Senator George McGovern of South Dakota, had become a candidate for President in 1968 after Robert Kennedy was assassinated, and was credited with keeping some of Kennedy's supporters engaged in practical politics. He endorsed his old neighbor Humphrey after the nomination and campaigned enthusiastically for him. Donald Fraser, member of Congress from Minneapolis, was vice-chairman and when McGovern became a 1972 Presidential candidate succeeded to the chairmanship.
71. See "Democrats Name Two Reform Groups," *New York Times* (February 9, 1969).
72. Commission on Party Structure and Delegate Selection to the Democratic National Committee, *Mandate for Reform* (Washington, D.C., April 1970).
73. These claims were fortified by a legal opinion by Democratic party general counsel Joseph Califano and by subsequent directives from national chairman Lawrence O'Brien. See Stearns, "The Presidential Nominating Process," pp. 36–38.
74. The full text of the guidelines is given in an appendix to this chapter.
75. *Mandate for Reform,* p. 34.
76. Shafer, "The Party Reformed," pp. 323–25.
77. *Ibid.,* pp. 324–25.

78. *Mandate for Reform,* p. 34.
79. See, however, for a thorough reconstruction, Shafer, "The Party Reformed."
80. *Cousins v. Wigoda* 419 *U.S.* 477 (1975). In this case the exclusion of duly certified delegates from Illinois by the credentials committee of the 1972 Democratic convention was upheld. The delegates had been slated by the Daley machine.
81. Excellent reviews of the field up to 1971 may be found in Alexander Heard, *The Costs of Democracy* (Chapel Hill: University of North Carolina Press, 1960), pp. 344–55; and Herbert E. Alexander, *Money in Politics* (Washington, D.C.: Public Affairs Press, 1972), pp. 183–251.
82. Frederick C. Mosher *et al., Watergate* (New York: Basic Books, 1974), p. 87.
83. See Nelson W. Polsby, "Policy Initiation in the American Political System," in I. L. Horowitz (ed.), *The Use and Abuse of Social Science* (New Brunswick, N.J.: Transaction Books, 1971), pp. 296–308; and Polsby, *Political Innovation in America: The Politics of Policy Initiation* (forthcoming: esp. Chap. 5).
84. A good part of this history remains to be written. For an overview see Frank J. Munger and Richard F. Fenno, Jr., *National Politics and Federal Aid to Education* (Syracuse, N.Y.: Syracuse University Press, 1962). See also Ted Bryant, "Carl Elliott: His Legislation Helped Millions Go to College," *Birmingham Post-Herald* (May 24, 1982).
85. P.L. 93–443 (October 15, 1974).
86. These limits were $1,000 for individuals, $5,000 for political committees, and $10,000 (per House candidate) and $20,000 (per Senate candidate) for political parties.
87. See P.L. 93-443, sec. 204.
88. The plaintiff first in alphabetical order was the conservative Senator from New York, James Buckley, who sued the Secretary of the Senate, Francis Valeo. Thus the case entered the law books as *Buckley et al. v. Valeo* 424 *U.S.* 1 (1976).
89. First amendment issues raised by this case are given an exceptionally clear and convincing exposition in Daniel D. Polsby, "*Buckley v. Valeo.* The Special Nature of Political Speech," *Supreme Court Review 1976* (Chicago: University of Chicago Press, 1977), pp. 1–44.
90. P.L. 94-283.
91. To $20,000 per year.
92. To $5,000 per year.
93. A good overview of the new law and its immediate impact from

several perspectives is Michael J. Malbin (ed.), *Parties, Interest Groups, and Campaign Finance Laws* (Washington, D.C.: American Enterprise Institute, 1980). See also Herbert E. Alexander and Brian A. Haggerty, *The Federal Election Campaign Act After a Decade of Political Reforms* (Los Angeles: Citizens' Research Foundation, 1981); and Gary R. Orren, "Presidential Campaign Finance: Its Impact and Future," *Commonsense* 4 (Number 2, 1981), pp. 50–66.

Chapter II: Consequences for Political Parties

1. See Everett Carll Ladd, Jr., " 'Reform' Is Wrecking the U.S. Party System," *Fortune* (November 1977), pp. 177–88; Jeane Jordan Kirkpatrick, *Dismantling the Parties: Reflections on Party Reform and Party Decomposition* (Washington, D.C.: American Enterprise Institute, 1979); Terry Sanford, *A Danger of Democracy* (Boulder, Colo.: Westview Press, 1981); Austin Ranney, *The Federalization of Presidential Primaries* (Washington, D.C.: American Enterprise Institute, 1978); Ranney, *Curing the Mischiefs of Faction* (Berkeley: University of California Press, 1975); Cyrus R. Vance, "Reforming the Electoral Reforms," *New York Times Magazine* (February 22, 1981), pp. 16, 62–69; James W. Ceaser, *Reforming the Reforms* (Cambridge, Mass.: Ballinger, 1982).
2. See William J. Crotty, *Decision for the Democrats* (Baltimore: Johns Hopkins University Press, 1978), pp. 254–73; Kenneth A. Bode and Carol F. Casey, "Party Reform: Revisionism Revised," in Robert A. Goldwin (ed.), *Political Parties in the Eighties* (Washington, D.C.: American Enterprise Institute, 1980), pp. 3–19; Charles Longley, "National Party Reform and the Presidential Primaries," prepared for delivery at the 1981 Annual Meeting of the Mid-West Political Science Association, April 15–19, 1981, Cincinnati. Bode was director of research of the McGovern Commission; Casey was a member of its research staff; and Longley was a summer intern on the staff of the Commission.
3. See John F. Bibby, "Party Renewal in the National Republican Party," in Gerald M. Pomper (ed.), *Party Renewal in America* (New York: Praeger, 1980), pp. 102–15.
4. McGovern campaign official Richard G. Stearns argued in 1972 that the shift in this burden was not intended by the McGovern Commission but arose as a consequence of a subsequent interpretation of guideline requirements in an exchange of letters "in late 1971 between [Democratic] party chairman Lawrence O'Brien and Rep. Donald Fraser of Minnesota, McGovern's successor as chairman of the reform commission. In these memoranda, Fraser out-

lined, and O'Brien assented to, a ruling that made a deficient percentage of minorities, women, or young persons *prima facie* evidence of discrimination, thereby making any delegation on which these groups were underrepresented susceptible to challenge . . . As a result, the convention was confronted with an unprecedented number of credentials challenges, many of them frivolous . . ." (Richard G. Stearns, "Reforming the Democrats' Reforms," *Washington Post Outlook,* December 3, 1972).

5. "Democratic Rules Reform Having Impact on '72 Race," *New York Times* (June 10, 1972). See also Apple, "Primaries: Change Is Profound," *New York Times* (February 10, 1972); Norman C. Miller, "Democrats' New Rules for Picking Delegates Add to '72 Uncertainty," *Wall Street Journal* (January 19, 1972).

6. Of course, in the case of Illinois this proved to be optimistic in 1972. See Stearns, "Reforming the Reforms."

7. Bode and Casey, "Party Reform: Revisionism Revised."

8. *Ibid.,* p. 17.

9. *Ibid.,* p. 10.

10. Report of the Commission on Presidential Nomination and Party Structure (Morley A. Winograd, chairman), *Openness, Participation, and Party Building: Reforms for a Stronger Democratic Party* (Washington, D.C.: Democratic National Committee, January 1978), p. 24.

11. Evidently subsidies do encourage candidacies. The creation of a state subsidy for gubernatorial election campaigns in New Jersey drew a record-breaking number of hopefuls out of the woodwork—twenty-one in all. See Larry Light, "Large Primary Field for New Jersey Governor," *Congressional Quarterly Weekly Report* (May 16, 1981), pp. 861–62.

12. F. Christopher Arterton, "Campaign Organizations Confront the Media-Political Environment," in James David Barber (ed.), *Race for the Presidency* (Englewood Cliffs, N.J.: Prentice-Hall, 1978), pp. 13–14.

13. "Democratic Rules Reform Having Impact on '72 Race," *New York Times* (June 10, 1972).

14. Systematic study of news coverage also shows a pattern highlighting early decisions, as the study based on 1980 data by Michael Robinson and Margaret Sheehan demonstrates. *Over the Wire and on TV* (New York: Russell Sage, forthcoming).

15. "TV's Newest Program: 'The Presidential Nominations Game,' " *Public Opinion* 1 (May–June 1978), pp. 41–46.

16. P.L. 93-443, Section 101 (b) (1): "Except as otherwise provided . . . no person shall make contributions to any candidate with

respect to any election for Federal office which, in the aggregate, exceed $1,000." See also Joel L. Fleishman, "Freedom of Speech and Equality of Political Opportunity: The Constitutionality of the Federal Election Campaign Act of 1971," *The North Carolina Law Review* 51 (January 1973), pp. 389–483; Fleishman, "Public Financing of Election Campaigns: Constitutional Constraints on Steps Toward Equality of Political Influence of Citizens," *ibid.,* 52 (December 1973), pp. 349–416; Michael Gartner, "Campaign Financing: A Dubious Law," *Wall Street Journal* (April 5, 1972).

17. Official Federal Election Commission figures for all 1980 prenomination presidential campaigns show the following distributions of sources of funding:

Table 2.3 Prenomination campaigns, 1980 sources of funding

	Percent of total contributions
Individual contributions	56.5%
Federal matching funds	23.5
Loans	12.3
Refunds	5.0
Political Action Committees	1.2
Other	1.5
	100.0%

	Individual contributions
Less than $500	55.4%
$500–$749	13.0
$750 or more	31.6
	100.0%

Source: Federal Election Commission, *FEC Reports on Financial Activity 1979–1980, Final Report, Presidential Prenomination Campaigns* (Washington, D.C.: Federal Election Commission, October 1981), pp. 1–5.

18. See James I. Lengle, *Representation and Presidential Primaries: The Democratic Party in the Post Reform Era* (Westport, Conn.: Greenwood, 1981), p. 83; Arlen J. Large, "All Those Candidates—Is It a Joke?," *Wall Street Journal* (January 28, 1972).

19. Mark Alan Siegel, "Toward a More Responsible Democratic Party: The Politics of Reform," Ph.D. dissertation, Northwestern University, June 1972, p. 237. Not until 1976, however, were all dele-

gates selected in the state primary. In 1968, 123 delegates had been selected in the June 18 primary, and 65 at-large delegates were selected at the June 28 state Democratic committee meeting. In 1972, 248 delegates were selected in the June 20 primary, and only 28 at-large delegates were selected at the June 24 state committee meeting. In 1976, all 274 delegates were selected in the April 6 state primary election (*Congressional Quarterly Weekly Report* (August 23, 1968), p. 225; (June 24, 1972), p. 1513; and (July 10, 1976), p. 1797).

20. Ranney, *The Federalization of Presidential Primaries,* pp. 2–3.
21. Defenders and critics of the contemporary nomination process are agreed on this point. See Everett Carll Ladd, Jr., *Where Have All the Voters Gone?* (New York: Norton, 1978); Crotty, *Decision for the Democrats;* Ranney, *The Federalization of Presidential Primaries.*
22. The compatibility of this discussion with David Truman's classic treatment of groups in the political process will no doubt be apparent. See Truman, *The Governmental Process* (New York: Knopf, 1951).
23. See Madison, *The Federalist #10* (New York: Modern Library, 1937), pp. 53–62. This is no place to quibble with James Madison, but it does seem that in *Federalist* 10 Madison does not foresee what has in fact occurred: that the constitutional machinery he and his colleagues devised mitigates the effects of faction by requiring factions to enter into coalitions.
24. See V. O. Key, Jr., *Southern Politics* (New York: Knopf, 1950), pp. 259–61; O. Douglas Weeks, "Texas: Land of Conservative Experience," in William C. Havard (ed.), *The Changing Politics of the South* (Baton Rouge: Louisiana State University Press, 1972), pp. 201–30, esp. pp. 212ff.; Eugene W. Jones, Joe E. Ericson, Lyle C. Brown, Robert S. Trotter, Jr., *Practicing Texas Politics* (Boston: Houghton Mifflin, 1974), pp. 74–81.
25. See Duane Lockard, *New England State Politics* (Princeton: Princeton University Press, 1959), pp. 228–319; Joseph I. Lieberman, *The Power Broker* (Boston: Houghton Mifflin, 1966); Joseph P. Lyford, *Candidate,* Eagleton Institute Case #9 (New York: McGraw-Hill, 1960).
26. These are not fictional numbers. Jimmy Carter won 22,875 votes in the 1976 New Hampshire primary, giving him 29.4% of the vote. Morris Udall was second with 23.9% of the vote, or 18,594 votes. One percent of the vote equaled 778 votes. In the 1972 Democratic primary, 1% of the vote equaled 888 votes, and in 1968, 550 votes. For the Republicans the numbers were somewhat

higher: 1040 votes equaled 1% in 1968, 1171 votes in 1972, and 1083 in 1976.

27. Here, for example, is a description of Jimmy Carter's 1976 strategy: "In the simplest terms, the strategy was to show well in the Iowa precinct caucuses on January 19, to stimulate press interest in Carter, and then to spend as much money as could be raised and borrowed—from friendly Atlanta banks and Georgia business interests—to win the New Hampshire and Florida primaries. Despite innumerable reports of Carter master plans and organizational genius, there was no plan and precious little organization after the Florida vote on March 9. The Carters gambled that their early victories would generate fundraising, organizational, and press 'momentum'—the political cliché used to describe what is happening when no one is sure," (Richard Reeves, *Convention* (New York: Harcourt Brace Jovanovich, 1977), p. 180).

Joel McCleary, Carter's national finance director told F. Christopher Arterton: "We had no structure after Florida; we had planned only for the short haul. After Florida, it was all NBC, CBS and the *New York Times*" (Arterton, "Campaign Organizations Confront the Media-Political Environment," in Barber (ed.), *Race for the Presidency,* p. 7).

Hamilton Jordan, in his famous August 4, 1974, memorandum laying out a strategic program for prospective candidate Jimmy Carter wrote:

The prospect of a crowded field coupled with the new proportional representation rule does not permit much flexibility. . . . No serious candidate will have the luxury of picking or choosing among the early primaries. To pursue such a strategy would cost that candidate delegate votes and increase the possibility of being lost in the crowd. I think we have to assume that everybody will be running in the first five or six primaries. A crowded field enhances the possibility of several inconclusive primaries with four or five candidates separated by only a few percentage points. Such a muddled picture will not continue for long as the press will begin to make "winners" of some and "losers" of others. The intense press coverage which naturally focuses on the early primaries plus the decent time intervals which separate the March and mid-April primaries dictates a serious effort in all of the first five primaries. Our "public" strategy would probably be that Florida was the first and real test of the Carter campaign and that New Hampshire would just be a warmup. In fact, a strong, surprise showing in New Hampshire should be our goal which would have tremendous impact on successive primaries. Our minimal goal in these early primaries would be to gain acceptance as a serious and viable candidate, demonstrate that Wallace is vulnerable and that Carter can appeal to the "Wallace" constituency, and show through our campaign a contrasting style and appeal. Our minimal goal would dictate at least a second-place showing in New Hampshire and Florida and respectable showings in Wisconsin, Rhode Island, and Illinois. Our national goals (which I think are highly attainable) would be to win New Hampshire and/or Florida outright, make strong showings in the other three early primary states and beat Wallace. [Martin Schram, *Running for President* (New York: Pocket Books, 1976), pp. 261–62.]

See also Jules Witcover, *Marathon* (New York: Viking Press, 1977). A strategy of differentiation and factional mobilization worked to Lyndon Johnson's advantage in his first election to the House of Representatives, in 1937, as Rowland Evans and Robert Novak describe in *Lyndon B. Johnson: The Exercise of Power* (New York: New American Library, 1966), pp. 7–8.

28. "Democratic Reforms: They Work," *Wall Street Journal* (May 16, 1972).

29. Jack W. Germond and Jules Witcover, *Blue Smoke and Mirrors* (New York: Viking, 1981), p. 96.

30. F. Christopher Arterton, "Campaign Organizations Confront the Media-Political Environment," in Barber (ed.), *Race for the Presidency*, p. 17.

31. Perhaps tradition or the herd insinct among journalists also plays a part; otherwise it is hard to explain the disparity in the coverage of the February 26 primaries in New Hampshire and Minnesota.

32. In 1980, for example, only 9 percent of all uncommitted delegates sent to the Democratic National Convention were selected in primary states and 62 percent of the uncommitted Republicans were from primary states. The numbers of uncommitteds in both cases were very small: 116.1 out of 3,311 Democratic delegates; 101 out of 1,994 Republicans. See Jonathan Moore (ed.), *The Campaign for President* (Cambridge, Mass.: Ballinger, 1981), p. 278.

33. For an early discussion of this phenomenon, see Stanley Kelley, *Professional Public Relations and Political Power* (Baltimore: Johns Hopkins University Press, 1956). See also, Larry J. Sabato, *The Rise of Political Consultants: New Ways of Winning Elections* (New York: Basic Books, 1981).

34. As Fred Barbash and Bill Curry, "Campaign '80: In FEC Maze, Auditors Supplant 'Fat Cats'," *Washington Post* (June 14, 1979) report:

At the George Bush for President headquarters in Houston, a 255-page manual of Federal Election Commission rules and regulations is called "The Bible." It holds all of man's accumulated knowledge on how to get eligible and stay eligible for federal matching presidential campaign funds.

If the manual is "The Bible," the Bush headquarters building itself is a temple to the FEC rules and regulations. The fund-raising operation is segregated on the fifth floor—because fund-raising expenses, including the rent, are to be segregated for FEC campaign expenditure reporting. The operation devoted to winning Texas '80 delegates to the 1980 Republican convention is confined to the third floor—because the state-by-state expenditures must be carefully controlled to avoid exceeding FEC limitations. The national political operation of the Bush campaign is handled separately on the second floor, to avoid commingling it with the other segregated expenditures. . . .

The campaigns have become more like businesses—centralized, heavily regulated, and loaded with accountants—with what many 1980 presidential campaign strategists foresee as a sharp decline in the potential for grass-roots participation and state-by-state flexibility. . . .

Officials from the various campaigns estimate that compliance—accountants to keep the records, computers to check them, copiers to copy them, and lawyers to interpret and (sometimes) get around them—will cost at least $1.5 million per campaign up to the time of the nomination conventions. "By the time the FEC comes in to audit us," said Stan Huckaby, head of an accounting firm that has been retained by Sen. Howard Baker's campaign, "we will have audited ourselves three times. There will be a manual check before it goes into the computer, a check by the computer and another check when it comes out.". . .

As a result, the accountants are likely to have expanded roles in campaign planning. "Before the law," said Huckaby, "the last person they wanted in a room before making a decision was an accountant. Now, the accountant is essential."

35. As Richard B. Cheney, who ran the Ford 1976 campaign rather sardonically put it, compliance with the law may mean a candidate is "better equipped to serve as director of the Office of Management and Budget than President" (Rhodes Cook, "Public Financing Alters Presidential Politics," *Congressional Quarterly Weekly Report* (October 6, 1979), p. 2228).

36. There is, for instance, the great example of Les Biffle, the Democratic secretary of the Senate in the 1950s, who went about the country during the 1948 campaign claiming he was a vacationing chicken farmer in order to take soundings of public opinion. Such a venture would today be regarded as unsystematic (*New York Times* (August 1, 1948), p. 49).

37. Joel Kotkin and Paul Grabowicz, "The New Star Trek: Cashing In on Politics," *Washington Post Outlook* (February 25, 1979):

Prominent among the emerging powers is a strange new breed of Hollywood movie stars and rock singers—not issue-advocates of the Robert Redford, Shirley MacLaine or Marlon Brando variety, but a group whose political talent is simply to raise more campaign cash in a night than others might lure in a year. Indeed, the new group fittingly cares far more about personalities than about issues, if it thinks about issues at all. . . . The growing trek of such stars to political campaigns derives, in large part, from 1974 changes in federal election laws that limited individual campaign contributions to $1,000. "The stars can contribute their services without being deemed as having gone over the $1,000 limit," observes Gray Davis, [Jerry] Brown's chief of staff and manager of his 1978 reelection. "Entertainers can attract contributions, particularly smaller ones, which would otherwise not be available to a candidate," Davis says. "People go to one of our concerts basically to see the Eagles perform. Frankly, we'd have trouble getting one-fifth the people there just to see Jerry.

Michele Willens, "The Hollywood-Washington Connection," *California Journal* (August 1979), pp. 265–68, reports:

In 1976, Jimmy Carter's lack-luster campaign was born again, thanks to a concert by the Allman Brothers. Jerry Brown's brief run for the presidency that same year was substantially funded—about 40 percent—by the entertainment industry.

See also Morton Mintz, "Defeat of PAC Restraints Could Cost Democrats the House in 1982," *Washington Post* (December 25, 1980); Dennis Farney, "A Modern Machine: How Savvy Matt Reese, a Political Consultant, Gets Out Winning Vote," *Wall Street Journal* (March 23, 1972); Theodore H. White, "The Making of the President Ain't What It Used To Be," *Life* (February 1980), pp. 66–80.

38. An example of these efforts would be the requirement that state delegations represent candidates who win even small amounts of support, as the Democrats put into their 1974 (Mikulski Commission) reforms. See David S. Broder, "New Democratic Rules: First Ballot Victory Unlikely," *Washington Post* (March 31, 1974).

39. See the comments of the professional political managers of a large number of 1980 candidates on this point in Jonathan Moore (ed.), *The Campaign for President*, pp. 1–50.

40. In 1980 the Democrats and Republicans together selected 5305 delegates of which exactly 217.1 came to either convention unpledged.

41. In 1976 the Republicans had a disagreement over pledged delegates and in 1980 repealed their requirement that delegates stay committed. Democrats are likely to do the same in 1984.

42. This may have cost the Democrats a seat in Congress at their next opportunity after the 1980 election. The first special Congressional election of the 97th Congress, occasioned by the death of Republican incumbent Tennyson Guyer, was held in June 1981 in the conservative 4th district of Ohio. Even though he was denied any help from the national Democratic party, as required by the resolution banning assistance to candidates refusing to back the ERA, the Democratic candidate, Dale Locker, came within 378 votes of winning the seat (Oxley, R, 41,904 (50.2%); Locker, D, 41,526 (49.8%)). Locker had voted against ERA as a state legislator, and personal convictions aside, must have known that advocating the ERA in the small towns of western Ohio would hurt

him more than national party aid could help. The resolution made it necessary for the party to withhold its aid regardless of the chance of victory.

43. In the Democratic case the rule is given in McGovern guideline 4(c): requiring state parties to "conduct the entire process of delegate selection in a timely manner, i.e., within the calendar year of the convention." In the Republican case the guideline is contained in Rule 31(o):

> No delegate shall be deemed eligible to participate in any district or state convention, the purpose of which is to elect delegates to the said national convention, who were elected prior to the date of issuance of the call of such national convention unless this rule be inconsistent with the law of the state.

The call is usually issued late in the year preceding the convention.

44. In *Cousins v. Wigoda* 419 *U.S.* 417 (1975) and *Democratic Party of U.S. v. La Follette* 67 *L. Ed. 2d* 82 (1981).

45. Richard B. Cheney, "The Law's Impact on Presidential and Congressional Election Campaigns," in Michael J. Malbin (ed.), *Parties, Interest Groups and Campaign Finance Laws* (Washington, D.C.: American Enterprise Institute, 1980), pp. 240–41. See also Bill Keller and Irwin B. Arieff, "As Campaign Costs Sky-Rocket, Lobbyists Take Growing Role in Washington Fund-Raisers," *Congressional Quarterly Weekly Report* (May 17, 1980), pp. 1333–46.

46. During the 1976 presidential election campaign the FEC spent much time and energy deciding three seemingly trivial cases. The judgment receiving the most publicity concerned a campaign button paid for by Rep. Edward Koch's (D-N.Y.) election committee. The button, which featured Koch's name along with Carter and Mondale, was feared by the committee to constitute a contribution to the Carter campaign. The FEC ruled, in a letter to the committee, that the button would not be considered such a contribution.

The FEC dealt with the same basic issue in two other cases. The FEC allowed Rep. Douglas Walgren (D-Pa.) to distribute a brochure containing a picture of the Congressman with Carter without counting it as a contribution to the Carter campaign. Representative Parren J. Mitchell (D-Md.) was not allowed to use an advertisement picturing him with Carter, Mondale, and Senate candidate Paul Sarbanes. The FEC published no reason for the different outcomes.

Michael J. Malbin, "After Surviving Its First Election Year, FEC Is Wary of the Future," *National Journal* (March 26, 1977), pp.

469–73. See also "The Koch Quandary and Other Matters," *ibid.*, p. 471.

47. P.L. 96-187. See Charles W. Hucker, "Congress Fine Tunes Campaign Law," *Congressional Quarterly Weekly Report* (January 5, 1980), p. 31. For an early assessment giving mixed notices, see Xandra Kayden, "Parties and the 1980 Presidential Election," in *Financing Presidential Campaigns* (mimeo), (Institute of Politics, John F. Kennedy School of Government, Harvard University, January 1982), Chapter 6.

48. A monographic treatment of the entire subject is Daniel A. Mazmanian, *Third Parties in Presidential Politics* (Washington, D.C.: Brookings Institution, 1974).

49. *Ibid.*, p. 5, and *Congressional Quarterly* provide useful data on third party presidential voting:

Table 2.7 Third parties in presidential elections receiving more than one percent of the popular vote, 1828–1980

Year	Party	Percentage of total votes cast
1832	Anti-Mason	7.8
1844	Liberty	2.3
1848	Free Soil	10.1
1852	Free Soil	4.9
1856	Whig-American	21.5
1860	Southern Democratic	18.1
	Constitutional Union	12.6
1880	Greenback	3.3
1884	Greenback	1.7
	Prohibition	1.5
1888	Prohibition	2.2
	Union Labor	1.3
1892	Populist	8.5
	Prohibition	2.3
1900	Prohibition	1.5
1904	Socialist	3.0
	Prohibition	1.9
1908	Socialist	2.8
	Prohibition	1.7
1912	Progressive*	27.4
	Socialist	6.0
1916	Socialist	3.2
	Prohibition	1.2
1920	Socialist	3.4

Table 2.7   (*continued*)

Year	Party	Percentage of total votes cast
1924	Progressive	16.6
1932	Socialist	2.2
1936	Union	2.0
1948	States' Rights Democratic	2.4
	Progressive	2.4
1968	American Independent	13.5
1972	American	1.4
1976	Independent (McCarthy)	1.0
1980	Independent (Anderson)	7.0
	Libertarian	1.0

*The vote for the Republican party was 23.2 percent, making the vote of the short-lived Progressive party the second highest of the election.

Sources: Robert A. Diamond (ed.), *Presidential Elections Since 1789* (Washington, D.C.: Congressional Quarterly, Inc., 1975), pp. 65–99; *Congressional Quarterly Weekly Report* (November 6, 1976), p. 3118; and *ibid.* (November 8, 1980), p. 3299. See also Daniel A. Mazmanian, *Third Parties in Presidential Politics* (Washington, D.C.: Brookings Institution, 1974).

50. This can be illustrated with figures for elected officeholders at the state level.

Table 2.8   Proportion of state officials that are either Democrats or Republicans

	Dem	Rep	State legislators % Major party	Others*	%
1959	4813	2510	96.7	250	3.3
1969	4284	3126	96.7	254	3.3
1977	5022	2348	98.3	127	1.7

	Dem	Rep	State executives % Major party	Others*	%
1959	339	96	89.7	50	10.3
1969	236	179	98.1	8	1.9
1977	242	123	97.9	8	2.1

*Other includes minor parties and non-partisan offfices

Sources: Council of State Governments, *The Book of the States,* Supplement I, 1959, Supplement I, 1969, *State Elective Officials and the Legislature, 1977.*

In some local jurisdictions, party designations are forbidden by law to candidates for public office.

51. See Douglas Rae, *The Political Consequences of Election Laws* (New Haven: Yale University Press, 1971).

52. P.L. 93-443, P.L. 94-283. See Stephen E. Gottlieb, "Putting Meaning into the Right of Association: The Job of Political Parties" (mimeo, presented at the American Bar Association's Wingspread Conference on the Presidential Selection Process, Racine, Wis., July 16–18, 1981), pp. 47–53.

53. P.L. 93-443, sec. 9008. See Gottlieb, "Putting Meaning into the Right of Association."

54. See Nicholas Zapple, "Historical Evolution of Section 315," in Austin Ranney (ed.), *The Past and Future of Presidential Debates* (Washington, D.C.: American Enterprise Institute, 1979), pp. 56–74.

55. See Joel Fleishman, "Freedom of Speech"; see also American Bar Association, Special Committee on Election Reform, *Symposium on Campaign Financing Regulation,* Tiburon, Calif., April 25–27, 1975, pp. 5–13; Daniel D. Polsby, *"Buckley v. Valeo:* The Special Nature of Political Speech," *Supreme Court Review* (1976), pp. 1–43; Frank Lynn, "Millionaires Drop Lobbying Unit; Change Laid to Fear of Publicity," *New York Times* (November 19, 1971).

56. On the Anderson campaign, see Dom Bonafede, "Can John Anderson Succeed Where Teddy Roosevelt Failed?" *National Journal* (May 17, 1980), pp. 806–10; Rhodes Cook, "High Hurdles for the Anderson Campaign," *Congressional Quarterly Weekly Report* (May 17, 1980), pp. 1315–18, and "Alternative Party Candidates May Have Substantial Impact on 1980 Presidential Election," *ibid.* (October 18, 1980), pp. 3143–49.

57. Table 2.9 compares the two parties in size.

Table 2.9 Trends in political affiliation: Size of the two parties

	Republican	*Democratic*	*Independent*
1937	34%	50%	16%
1940	38	42	20
1944	39	41	20
1949	32	48	20
1952	34	41	25
1960	30	47	23
1964	25	53	22

212   *Notes*

Table 2.9 (*continued*)

	Republican	Democratic	Independent
1968	27	46	27
1972	28	43	29
1975	22	45	33
1980 (Apr–Jun)	23	47	30
1980 (Oct–Dec)	26	43	31

Source: *The Gallup Opinion Index,* Report No. 183 (December 1980), p. 64.

On the issue of cohesion there are considerable data comparing
Democrats and Republicans on a variety of issues but a full-blown
examination of the proposition that Democrats are, on the whole,
more dispersed in their political attitudes than Republicans does
not seem to have been conducted.

One simple measure of dispersion, the standard deviation, when
applied to Democratic and Republican respondents to the 1972
University of Michigan Center for Political Studies election study
on twelve issues spanning a wide gamut of concerns showed Democrats more dispersed on all twelve, as Table 2.10 indicates:

Table 2.10 Dispersion of attitudes comparing Democrats and
Republicans across a seven-point scale, 1972

Issue	Standard deviation	
	Democrats	Republicans
1. Liberal/Conservative Self-Identification	1.281	1.134
2. Government Aid to Minorities	2.079	1.752
3. Rights of the Accused	2.169	2.011
4. Women's Equality	2.286	2.227
5. Government Fight Air Pollution	1.921	1.879
6. Government Health Insurance	2.445	2.291
7. Busing for Integration	1.893	1.196
8. Legalize Marijuana	2.249	1.989
9. Government Action on Inflation	1.549	1.492
10. Vietnam Withdrawal	1.922	1.798
11. Progressive Income Tax	2.390	2.231
12. Government Job Guarantees	2.064	1.818

Source: Center for Political Studies, University of Michigan

The differences in the numbers do not look enormous, but it should be remembered that the dispersion of the entire populations of Democrats and Republicans can only occur over a range of seven points for each, and consequently small differences in standard deviations do signify differences that deserve to be taken seriously.

Rough confirmation of the overall conclusion is provided by a CBS News/*New York Times* survey in the spring of 1981 on ten different issues. Here rank and file Democrats were less dispersed than Republicans on four issues (Equal Rights Amendment, Allow Abortions with Doctor's Consent, Government Spending on Job Programs, Aid to the Arts) and more dispersed on six (Government Spending on Military-Defense, and on Food Stamps, Reagan's Tax Cut, Prayer Amendment, Government Regulation of Business). See Martin Plissner and Warren Mitofsky, "Political Elites," *Public Opinion*, October/November 1981, 49.

In general, dispersion among Democrats, cohesion among Republicans, appear to be what politicians and journalists have commonly believed about the two parties over the last generation or so. An example of this sort of sentiment is expressed in Dean Acheson, *A Democrat Looks at His Party* (New York: Harper, 1955), pp. 23–27:

> From the very beginning, the Democratic party has been broadly based . . . It has been the party of the many . . . They have many interests, many points of view, many purposes to accomplish . . . It is this multiplicity of interests which . . . is the principal clue in understanding the vitality and endurance of the Democratic party. The economic base and the principal interest of the Republican party is business . . . Here lies the significant difference between the parties, the single-interest party against the many-interest party.

This conclusion in its general outline does not seem to be a matter of much partisan controversy. John Bibby, an active Republican political scientist, has recently referred to the "intense ideological faction within the more ideologically homogeneous GOP, the conservatives" ("Party Renewal in the National Republican Party," in Pomper (ed.), *Party Renewal in America*, p. 106), and Charles O. Jones, in his excellent book *The Republican Party in American Politics* (New York: Macmillan, 1965), chafes only slightly at the party-of-business label (p. 3) while firmly identifying the Republicans as the smaller, minority party. See also Gallup Report no. 194 (November 1981), pp. 29–45, which shows Democrats greatly outnumbering Republicans (45 percent to 27 percent). The Democratic party is also identified by Gallup survey respondents as favoring farmers, skilled workers, unskilled workers, labor union members, blacks, small business people, women, unemployed people, young people, retired people and "the average citizen." The Republicans were identified as better for upper income people, corporation executives, professional and business people, and (narrowly) white collar workers.

The plausibility of the claim that the two parties are indeed different is greatly strengthened by elite studies such as David R. Mayhew's *Party Loyalty Among Congressmen: The Difference Between Democrats and Republicans 1947–1962* (Cambridge, Mass: Harvard University Press, 1966). Mayhew's conclusion is that the Republicans in the House during the period he studied adhered to a policy of ideological cohesion and commitment to conservative principles, voting against their party colleagues who pursued purely district interests on individual issues. The Democrats, on the other hand, voted in favor of the entire range of particular interests put forward by their various factions, and maintained a broadly "inclusive" coalition.

Table 2.11 gives recent information on the House of Representatives, and shows Democrats far more dispersed across the available ideological alternatives than Republicans:

Table 2.11   Ideological divisions in the House, 86th Congress, 1st Session (1959) and 95th Congress, 1st Session (1977)

	Democrats		Republicans	
86th Congress				
Liberals	51%	(142)	4%	(6)
Moderates	13	(37)	15	(23)
Conservatives	36	(102)	81	(124)
95th Congress				
Liberals	50%	(145)	1%	(2)
Moderates	30	(86)	20	(29)
Conservatives	20	(57)	79	(115)

Source: Lawrence C. Dodd and Bruce I. Oppenheimer, "The House in Transition: Change and Consolidation," in Dodd and Oppenheimer (eds.), *Congress Reconsidered* (Washington, D.C.: Congressional Quarterly, 1981), p. 35.

Finally, a recent study of political elites by CBS News shows a less dramatic, but generally consistent pattern:

Table 2.12 Distributions of Democrats and Republicans by ideology

	Rank and file		National committee members		1980 National convention delegates	
	Dem.	Rep.	Dem.	Rep.	Dem.	Rep.
Liberal	24%	11%	36%	1%	46%	2%
Moderate	42	33	51	31	42	36
Conservative	29	51	4	63	6	58

Source: Martin Plissner and Warren Mitofsky, "Political Elites," *Public Opinion* (October/November 1981), pp. 47–48.

On the ten issues covered in the CBS poll, Democratic National Committee members were more dispersed in their opinions than Republican National Committee members on all issues except the School Prayer Amendment, E.R.A., and Abortions with Doctor's Consent (Plissner and Mitofsky, Table III, p. 49).

58. See Hugh A. Bone, *Party Committees and National Politics* (Seattle: University of Washington Press, 1958); Cornelius P. Cotter and Bernard C. Hennessy, *Politics Without Power* (New York: Atherton Press, 1964); John F. Bibby, "Party Renewal in the National Republican Party," in Pomper (ed.), *Party Renewal in America;* and Cornelius P. Cotter and John F. Bibby, "Institutional Development of Parties and the Thesis of Party Decline," *Political Science Quarterly* 95 (Spring 1980), pp. 1–27.

59. See Thomas E. Mann and Norman J. Ornstein, "The Republican Surge in Congress," in Ranney (ed.), *The American Elections of 1980,* pp. 263–302, esp. pp. 263–67; John F. Bibby, "Party Renewal in the National Republican Party," in Pomper (ed.), *Party Renewal in America;* Cotter and Bibby, "Institutional Development of Parties and the Thesis of Party Decline," esp. pp. 18–19.

60. Compare, for evidence, the disparities between party identifications as monitored by public opinion surveys and party presidential vote:

Table 2.13   Party identification compared with party
presidential vote

	Dem. % of party ID	Dem. % of pres. vote	Difference	Rep. % of party ID	Rep. % of pres. vote	Difference
1968	55	43	−12	33	43	+10
1972	51	38	−13	34	61	+27
1976	52	50	−02	33	48	+15
1980	52	41	−11	33	51	+18

Source: Center for Political Studies, University of Michigan. (Independent leaners included with the party toward which they lean.)

61. This notion is argued at length in Nelson W. Polsby, "What If Robert Kennedy Had Not Been Assassinated in 1968," in Polsby (ed.), *What If? Explorations in Social Science Fiction* (Lexington, Mass: Lewis Publishing Co., forthcoming).

62. Much evidence on this point can be found in Jeane Kirkpatrick, "Representation in the American National Conventions: The Case of 1972," *British Journal of Political Science* 5 (July 1975), pp. 265–322. In 1972, even delegates to the Republican National Convention were more representative of the views of rank-and-file Democrats than were McGovern delegates to the Democratic convention (Tables 9–10, pp. 304–5).

63. See Table 2.9 above.

64. On the reasons for defection: one year in recent history, 1952, seems to have featured more Republican "pull" than Democratic "push," for defecting Democrats. See Herbert H. Hyman and Paul B. Sheatsley, "The Political Appeal of President Eisenhower," *Public Opinion Quarterly* 1 (Winter 1955–56), pp. 26–39. For the elections since 1968, most defecting Democrats appear to have been more at odds with the Democratic candidate than pleased with the Republican alternative. For 1972, see Jeane Kirkpatrick, "Representation in the American National Conventions: The Case of 1972"; Arthur H. Miller, Warren E. Miller, Alden S. Raine, and Thad A. Brown, "A Majority Party in Disarray: Policy Polarization in the 1972 Election," *American Political Science Review* 70 (September 1976), pp. 73–78; George Gallup, "Democratic Defections Expected To Shatter Voting Patterns," *Washington Post* (October 19, 1972); Haynes Johnson, "Vital Bloc Breaks Democratic Ties," *Washington Post* (October 8, 1972); Richard Harwood, " 'Ethnics' Alienated by Forces of Change," *Washington Post* (October 9, 1972).

For 1980, see Adam Clymer, "The Democrats Look For New Ideas, and Jobs," *New York Times* (November 9, 1980); Clymer, "Displeasure with Carter Turned Many to Reagan," *ibid.;* and especially William Schneider, "The November 4 Vote for President: What Did It Mean?" in Ranney (ed.), *The American Elections of 1980,* pp. 212–62.

65. See Everett Carll Ladd, Jr., "The Brittle Mandate: Electoral Dealignment and the 1980 Presidential Election," *Political Science Quarterly* 96 (Spring 1981), pp. 1–25; also Adam Clymer, "Displeasure with Carter Turned Many to Reagan"; and data in *Public Opinion* (December/January 1981), p. 43; Arthur Miller, "What Mandate? What Realignment?" *Washington Post Outlook* (June 28, 1981); Louis Harris, "No Mandate for a Switch on Social Questions Seen," *Washington Post* (December 4, 1980); George Skelton, "Conservative Mandate for Reagan Contains Limits," *Los Angeles Times* (November 20, 1980); *Public Opinion* (December/January 1981), pp. 24–25.

66. See Table 2.9 above.

67. In 1980, as in the previous fifteen congressional elections, more people voted for Democratic than for Republican House candidates. The popular vote for House candidates in 1980 was 51 percent Democratic, 48 percent Republican (*Public Opinion* (December/January), p. 24). Even after the 1980 elections, Democrats controlled almost twice the number of state legislative houses nationwide (63/34), and far more Democrats than Republicans sat as members of state legislatures. The figures are: Democratic members, 4,497; Republican members, 2,918 (*ibid.,* p. 25). And although the Republicans won more senatorial elections in 1980 this result was achieved without a Republican landslide in the popular vote. Without counting votes cast for the unopposed Democratic Senator from Louisiana, Russell Long, almost 3 million more people voted for Democratic candidates for the Senate than voted for Republicans. Republicans won 21 out of the 33 Senate seats up for a decision in 1980, but mostly by small margins. Of the fifteen races where the winner won by the biggest margins, nine were won by Democrats. See Nelson W. Polsby, "Party Realignment in the 1980 Election," *Yale Review* (Fall 1982).

68. Chicago Mayor Richard J. Daley, who had been ejected as a delegate from the Democratic convention of 1972, even though he and his slate had won their places in the Illinois primary, gave his last hurrah in 1976 when far in advance of the convention he threw in the towel and endorsed Jimmy Carter. On June 8, 1976,

when Carter had 38 percent of the delegates then allocated, Mayor Daley said, "This man, Carter, has fought in every primary, and if he wins in Ohio, he'll walk in under his own power." See Jules Witcover, *Marathon*, p. 349. Carter won in Ohio but on the same day lost in New Jersey and California. The final primary elections of June 9 gave Carter 39 percent of the total delegates. The remaining 61 percent were widely spread, including an uncommitted 18 percent. See Donald R. Matthews, "Winnowing," in J. D. Barber (ed.), *Race for the Presidency*, p. 72, Table 3.

Chapter III: Consequences for Governing

1. Two good overviews giving twentieth-century historical background on cabinet selection up to the Eisenhower administration are Richard F. Fenno, Jr., *The President's Cabinet* (Cambridge, Mass.: Harvard University Press, 1959) and Laurin L. Henry, *Presidential Transitions* (Washington, D.C.: Brookings Institution, 1960).

2. Neither, in Carter's case, did his campaign rhetoric. See Jack Knott and Aaron Wildavsky, "Jimmy Carter's Theory of Governing," *The Wilson Quarterly* 1 (Winter 1977), pp. 49–67, for an excellent preview of the concerns of the Carter presidency based on his campaign rhetoric.

3. For an account of the collective activities of the cabinet in Eisenhower's first term see Robert L. Donovan, *Eisenhower: The Inside Story* (New York: Harper, 1956).

4. Douglass Cater, "A New Style, a New Tempo," *The Reporter* (March 16, 1961), pp. 28–30.

5. A number of important issues are involved in the appointment of an attorney general, including the management of criminal indictments, appointments to the judiciary, and oversight of the activities of the FBI. Some of these issues cut unusually close to the political interests of Presidents, a fact that was not lost on Kennedy so soon after the difficulties, small-scale scandals, and embarrassments that had necessitated an appointment of a special prosecutor and the replacement of one attorney general in the Truman administration. See Cabell Phillips, *The Truman Presidency* (New York: Macmillan, 1966), pp. 403–14; and Jules Abels, *The Truman Scandals* (Chicago: Henry Regnery Co., 1956). For an account of Robert Kennedy's performance as attorney general see Victor Navasky, *Kennedy Justice* (New York: Atheneum, 1971). Other relevant works on John Kennedy and his cabinet include Adam Yarmolinsky, "The Kennedy Talent Hunt," *The Reporter*

(June 8, 1961), pp. 22–25, and Richard F. Fenno, Jr., "The Cabinet: Index to the Kennedy Way," *New York Times Magazine* (April 23, 1962), pp. 13ff.

6. An account of President Nixon's administrative goals and activities early in his Presidency is contained in Rowland Evans and Robert Novak, *Nixon in the White House: The Frustration of Power* (New York: Random House, 1971). It is useful to contrast Richard Nathan's *The Plot That Failed: Nixon and the Administrative Presidency* (New York: John Wiley, 1975), written from the perspective of later events.

7. On Nixon's first reorganization, see Evans and Novak, *Nixon in the White House,* pp. 237–41.

8. On Nixon and labor, see the following more or less representative news stories: Richard J. Levine, "Labor Taking a Tougher Line with Nixon," *Wall Street Journal* (February 20, 1970); Byron E. Calame, "Nixon-Labor Rift: Back to Normal," *ibid.* (August 24, 1971); "Nixon and the Unions: Can the 'Honeymoon' Last?," *U.S. News & World Report* (April 21, 1969), pp. 82–84; "Unions Open Fire on Nixon over Jobs, Civil Rights," *ibid.* (March 9, 1970), pp. 69–70; "George Meany Cuts His Ties with the White House," *ibid.* (September 10, 1973), pp. 78–79; "Meany Steps Up His War on Richard Nixon," *ibid.* (October 29, 1973), pp. 90–91; James M. Naughton, "President Asks Labor's Support; Reception Is Cool," *New York Times* (November 20, 1971); Philip Shabecoff, "Leader of Labor Declares Nixon Sought Incident," *ibid.* (November 23, 1971); Shabecoff, "Union Chiefs See New Meany Role," *ibid.* (November 24, 1971); Naughton, "Nixon Focusing on Meany as Likely Campaign Target," *ibid.* (November 25, 1971); George Meany, "Labor and Nixon," *The American Federationist* (December 1971), pp. 2–8; Meany, "A Dark Shadow of Shame over the Spirit of America," *ibid.* (November 1973), pp. 2–6. See also Peter Brennan's *Current Biography* entry, 1973, pp. 59–62.

9. On Nixon and the scientific community, see, for example, Daniel S. Greenberg, "Science Under Nixon: Influence Has Declined in National Affairs," *Science* 169 (September 11, 1970), pp. 1056–57. Greenberg says (p. 1056): "The scientific presence in Washington that grew up after World War II was never so potent as alarmed political traditionalists made it out to be; nor was it ever so unheeded and abused as many scientists made it out to be. But a look into science's Washington outposts after 2 years' absence quickly confirmed my impression that, however powerful the community may once have been in national affairs, 20 months under Nixon have inflicted upon it a gigantic loss of influence,

visibility, and confidence. . . . [T]here is little to suggest that the President accords scientific activity any special or privileged role in national life, and there is a good deal to suggest that the President, as well as many of his closest advisers, regard the scientific community as having succeeded in making unwarranted claims on national resources and political sympathy." See also "Nixon on Science," *ibid.* 174 (October 29, 1971), p. 477; "White House Presents Vapid Technology Plan," *ibid.* 175 (March 24, 1972), p. 1343; "Congress Holds Down NSF Budget; Nixon Vetoes HEW Bill," *ibid.* 177 (September 1, 1972), p. 775; "OMB: Hand in the Till or on the Tiller?," *ibid.* 179 (March 2, 1973), p. 879; and Deborah Shapley, "White House Science: Hail and Farewell," *ibid.* 179 (March 30, 1973). On Nixon and education, see "Battle Looms over Funds for Education," *The Chronicle of Higher Education* 4 (January 26, 1970); and Cheryl M. Fields, "Nixon's Legislative Plans Called Little Help to Colleges," *ibid.* 4 (May 18, 1970).

10. The tip of the iceberg was revealed in a story in the normally prosaic, civil service routine-oriented *Federal Times,* Inderjit Badhwan, "Government-wide Patronage Deals" (September 25, 1974). See also *Documents Relating to Political Influence in Personnel Actions at the Department of Housing and Urban Development,* published by the Subcommittee on Manpower and Civil Service of the House of Representatives Committee on Post Office and Civil Service, December 12, 1974 (Washington, D.C.: U.S. Government Printing Office, 1974), and Dom Bonafede, "Nixon Personnel Staff Works To Restructure Federal Policies," *National Journal* (November 12, 1971), pp. 2440–48. On July 19, 1973, the *Washington Post* published a list of 94 persons formerly employed in Mr. Nixon's 1972 campaign effort or on the White House staff who had been deployed out into cabinet departments or regulatory agencies. This was characterized as a "partial list." Additional discussion of the public administration problems of the Nixon Presidency together with documentation, is contained in Subcommittee on Manpower and Civil Service, *Final Report on Violations and Abuses of Merit Principles in Federal Employment* (Washington, D.C.: U.S. House of Representatives Committee on Post Office and Civil Service, December 30, 1976). See also Richard Nathan, *The Plot That Failed.*

11. For a famous discussion of the theory underlying this view see Woodrow Wilson, *Congressional Government* (New York: Meridian, 1956; first published 1885).

12. See Lou Cannon, "OEO Chief Savors Shutdown," *Washington Post* (February 4, 1973); Jules Witcover, "OEO Dismantlers Proceed

with Speed, Zeal: Fear, Rumors Plague Staff," *Washington Post* (February 17, 1973); and Austin Scott, "Plans To Neutralize Hill Revealed," *ibid.*

13. See in general, Richard P. Nathan, Allen D. Manvel, Susannah E. Calkins, and associates, *Monitoring Revenue Sharing* (Washington, D.C.: Brookings Institution, 1975); and Charles L. Schultze, Edward R. Fried, Alice M. Rivlin, and Nancy H. Teeters, *Setting National Priorities: The 1972 Budget* (Washington, D.C.: Brookings Institution, 1971), pp. 134–57.

14. ". . . the growth of the Executive Office of the President was relatively constant, until the late 1960s when the annual growth rate doubled; . . . with the advent of the Nixon Administration the growth rate increased almost 400% over the last part of the Johnson Administration" (U.S. House of Representatives, Committee on Post Office and Civil Service, *A Report on the Growth of the Executive Office of the President, 1955–1973* (Washington, D.C.: Government Printing Office, 1972), p. 3). See also Evans and Novak, *Nixon in the White House*, pp. 237–41. On impoundment, see Louis Fisher, *Presidential Spending Power* (Princeton: Princeton University Press, 1975), pp. 147–74, 175–201. In a January 31, 1973, news conference, Nixon claimed "the constitutional right for the President of the United States to impound funds . . . is absolutely clear . . . I will not spend money if the Congress overspends, and I will not be for programs that will raise the taxes and put a bigger burden on the already overburdened American taxpayer" ("Transcript of the President's News Conference on Foreign and Domestic Matters," *New York Times*, February 1, 1973).

15. The press release proclaiming the reorganization is dated January 5, 1973, to coincide with the beginning of Mr. Nixon's second term. Useful background material is contained in *Papers Relating to the President's Departmental Reorganization Program: A Reference Compilation* (Washington, D.C.: Government Printing Office, 1972). The reorganization plan never took effect.

16. For an expansion of this point, see Nelson W. Polsby, *Political Promises* (New York: Oxford University Press, 1974), pp. 6–14.

17. As the author of the *Federalist* 51—either Hamilton or Madison—put the matter: "To what expedient, then, shall we finally resort, for maintaining in practice the necessary partition of power among the several departments, as laid down in the Constitution? . . . [B]y so contriving the interior structure of the government as that its several constituent parts may, by their mutual relations, be the means of keeping each other in their proper places." For more

modern accounts see Richard E. Neustadt, *Presidential Power* (New York: Wiley, 1980); E. Pendleton Herring, *The Politics of Democracy* (New York: Holt, 1940); and Herbert Agar, *The Price of Union* (Boston: Houghton Mifflin, 1950).

18. Leader of this fashion, as of so many, was Arthur Schlesinger, Jr., *The Imperial Presidency* (Boston: Houghton Mifflin, 1973), p. 405. See also Theodore Sorenson, *Watchmen in the Night* (Cambridge, Mass.: MIT Press, 1975).

19. See Herbert Y. Schandler, *The Unmaking of a Presidency: Lyndon Johnson and Vietnam* (Princeton, N.J.: Princeton University Press, 1977), pp. 256–65, and Clark Clifford, "Vietnam Reappraised: The Personal History of One Man's View and How It Evolved," *Foreign Affairs* 47 (July 1969), pp. 601–22.

20. Perhaps the most remarkable modern exception is Lawrence O'Brien of Springfield, Massachusetts, who began his Washington career as John F. Kennedy's chief of legislative liaison and stayed on to become postmaster general under Lyndon Johnson and do two stints as chairman of the Democratic National Committee. His metamorphosis from JFK's patronage assistant and ambassador to Capitol Hill to Washington careerist got him in great trouble with other Kennedy loyalists. O'Brien says: "My decision to stay [in the Johnson White House] came at about the time that . . . others of the Kennedy team were leaving the government, and there was some bitterness at my staying. . . . I knew that some of them hated my guts for staying with Lyndon Johnson" (Lawrence F. O'Brien, *No Final Victories* (Garden City, N.Y.: Doubleday, 1974), p. 180).

21. Safire, *Before the Fall* (Garden City, N.Y.: Doubleday, 1975), pp. 250–62.

22. John Pierson, "The Job Is Demanding, But the Little Extras Ease a President's Lot," *Wall Street Journal* (April 4, 1969).

23. Bob Woodward and Carl Bernstein, *The Final Days* (New York: Simon and Schuster, 1976), p. 344, describe a scene in the White House on August 2, 1974: "The air conditioner was on high and, as usual, a fire was burning in the fireplace." Washington summers are not noted for their brisk weather.

24. See Richard E. Neustadt, *Presidential Power* (New York: John Wiley, 1980).

25. See Raymond E. Wolfinger, "Why Political Machines Have Not Withered Away and Other Revisionist Thoughts," *Journal of Politics* 34 (May 1972), pp. 365–98; Milton L. Rakove, *Don't Make No Waves—Don't Back No Losers* (Bloomington: Indiana University Press, 1975); and Fred I. Greenstein, *The American Party System and*

the American People (Englewood Cliffs, N.J.: Prentice-Hall, 1963).

26. Richard F. Fenno, Jr., *The President's Cabinet* (Cambridge, Mass.: Harvard University Press, 1959), is especially good at exploring the problem of competence versus other values in cabinet formation.

27. Biographical material on the Carter cabinet can be found in the *Congressional Quarterly* publication *President Carter* (Washington, D.C.: Congressional Quarterly, April 1977), pp. 22–43, and in *Current Biography* under alphabetical listings for 1977, except for Zbigniew Brzezinski, 1970, pp. 53–55; Patricia Harris, 1965, pp. 189–91; James Schlesinger, 1973, pp. 379–81; Charles L. Schultze, 1970, pp. 379–81; and Stansfield Turner, 1978, pp. 431–34. General commentary on President Carter's cabinet-building activity includes David Cohen, "Carter Has a Clear Field in Filling Jobs," *Los Angeles Times* (November 12, 1976); Karen Elliott House, "Seek and Ye Shall Find: Well Maybe Not So in Carter's Regime," *Wall Street Journal* (February 2, 1977); Alan L. Otten, "Politics & People: First Notices," *Wall Street Journal* (February 17, 1977); Leslie H. Gelb, "Carter Finding Few Outsiders," *New York Times* (December 15, 1976); Helen Dewar and Hobart Rowen, "Carter in No Hurry on Cabinet Selections," *Washington Post* (December 5, 1976); and Bruce Adams and Kathryn Kavanagh-Baran, *Promise and Performance* (Lexington, Mass.: Lexington Books, 1979).

28. See Laurence Stern, "Young: 'Point Man' for New Policy," *Washington Post* (March 13, 1977); William Greider, "Trilateralists To Abound in Carter's White House," *Washington Post* (January 16, 1977); Alan Baron, "Special Report: Carter's Foreign Policy Appointments," *The Baron Report* (January 25, 1977).

29. Bergland grew up on a Minnesota farm and graduated from the University of Minnesota School of Agriculture. He was a field representative for the Minnesota Farmers' Union before buying a farm of his own in 1950. During the 1960s he was a regional official of the Department of Agriculture's Agricultural Stabilization and Conservation Service. In 1970 he was elected to Congress and served on the House Agriculture Committee.

30. See Margot Hornblower, "A New Breed Shakes Old Order at Interior Dept.," *Washington Post* (April 3, 1977).

31. President Carter seems to have been determined to give this particular position to a woman. Mrs. Kreps was appointed after Jane Cahill Pfeiffer turned the job down.

32. Patricia Roberts Harris, Brock Adams, and Joseph Califano.

33. See Joseph Califano, *Governing America* (New York: Simon and

Schuster, 1981). Mrs. Harris replaced Califano as Secretary of Health, Education and Welfare. Califano quotes President Carter at the time of the shake-up that removed him from office and sent Mrs. Harris to HEW as telling a cabinet meeting to evaluate all the assistant secretaries in their departments and "get rid of all those who are incompetent except minorities and women" (p. 431). See also Herman Nickel, "Carter's Cactus Flower at HUD," *Fortune* (November 6, 1978), pp. 110–13.

34. Bert Lance, budget director; Griffin Bell, attorney general; and Andrew Young, ambassador to the U.N. Each returned to Georgia following his tour of duty with the Carter administration.

35. Admiral Stansfield Turner, CIA director.

36. See Jack Nelson, "Cabinet Choices Disappoint Some of Carter's Backers," *Los Angeles Times* (December 26, 1976); Edward Walsh, "Costanza Defends Carter on Recruitment of Women," *Washington Post* (January 28, 1977); Fred L. Zimmerman, "Carter's Cabinet: A Johnsonian Mix of Old Faces," *Wall Street Journal* (December 27, 1976); Edward Walsh, "Financial Data Are Released for Cabinet, Four Aides," *Washington Post* (February 26, 1977); David S. Broder, "No New 'Generation of Leaders'," *Washington Post* (December 24, 1976); Roger Morris, "Jimmy Carter's Ruling Class," *Harper's* (October 1977); Adams and Kavanagh-Baran, *Promise and Performance*.

37. As Robert Axelrod demonstrated in his "1976 Update," *American Political Science Review* 72 (June 1978), pp. 622–24, to his valuable "Where the Votes Come From" series, which began in the *American Political Science Review* 66 (March 1972), pp. 11–20. Axelrod says (p. 622), "For the Democrats, the New Deal coalition made a comeback in 1976. . . . The Democrats got a majority of the votes from each of the six diverse minorities which make up their traditional coalition: the poor, blacks, union families, Catholics, southerners, and city dwellers."

38. See Elizabeth Drew, "Our Far-Flung Correspondents: Settling In," *The New Yorker* (February 28, 1977), pp. 82–88.

39. News accounts of the extraordinary relations between President Carter and Congress are legion, and can be found for each year of his incumbency, as this sampling will attest: *1976:* Richard Harwood, "Gaffes, Strategy Errors Whittled Carter's Big Lead," *Washington Post* (November 4, 1976); *1977:* Martin Tolchin, "Byrd, Hinting Strained Relations, Says Carter Fails To Seek Advice," *New York Times* (January 27, 1977); Elizabeth Drew, "Settling In"; Dennis Farney, "Jimmy Carter and the Insiders," *Wall Street Jour-*

nal (April 11, 1977); Norman C. Miller, "Carter Not Playing by the Unwritten Rules of the Game," *Wall Street Journal* (April 22, 1977); Adam Clymer, "Carter's Woes with Congress," *New York Times* (June 2, 1977); Graham Hovey, "Congress and Foreign Policy: Snags in Both Houses Indicate Administration Is Not Handling Legislative Relations Properly," *New York Times* (June 22, 1977); *1978:* "His Mood at Mid-Term," *Newsweek* (August 28, 1978), pp. 20–21. *1979:* Meg Greenfield, "What Carter Thinks He's Doing," *Newsweek* (February 26, 1979), p. 100; Steven V. Roberts, "Carter, at Midterm, Is Still 'Outsider' to Many in Congress," *New York Times* (March 7, 1979); Martin Schram, "Carter's Chief Opponent in '80 Campaign May Be Congress," *Washington Post* (May 21, 1979); Steven V. Roberts, "Carter and the Congress: Doubt and Distrust Prevail," *New York Times* (August 5, 1979); Martin Tolchin, "Slow Improvement Is Seen in White House Relations with Congress," *New York Times* (November 26, 1979); *1980:* Jack Anderson, "The Gang That Thought Mineta Was Italian," *Washington Post* (May 25, 1980); Albert R. Hunt, "Carter's Congressional Record," *Wall Street Journal* (September 2, 1980); Irwin B. Arieff, "President Remains an Outsider: Carter-Congress Relations Still Strained Despite Gains," *Congressional Quarterly* (October 11, 1980), pp. 3095–3102.

The tenor of these public stories was amply confirmed in private conversation by members of Congress, notably by members ideologically in overwhelming agreement with the President. An excellent overview of Jimmy Carter's difficulties with the Washington community and especially Congress is Haynes Johnson's *In the Absence of Power* (New York: Viking, 1980). See also Betty Glad, *Jimmy Carter: In Search of the Great White House* (New York: Norton, 1980), for example, p. 420.

40. As the authoritative *CQ Almanac* said:

[T]he first session of the 89th, starting early and working late, . . . passed more major legislation than most Congresses pass in two sessions. The scope of the legislation was even more impressive than the number of major new laws. Measures which, taken alone, would have crowned the achievements of any Congress, were enacted in a seemingly endless stream.

The pace of the session was so breathless as to cause a major revision of the image, widely prevalent in preceding years, of Congress as structurally incapable of swift decision, prone to frustrate demands for progress. The change was due to three primary factors not always present in past years: The decisive Democratic majorities elected in 1964, the personal leadership of President Johnson, and the shaping of legislation to obtain maximum political support in Congress.

"Congress 1965—The Year in Review," *Congressional Quarterly Almanac* 21 (Washington, D.C.: Congressional Quarterly, 1965), p. 65.

41. See Norman J. Ornstein, Robert L. Peabody, and David W. Rohde, "The Contemporary Senate: Into the 1980s," in Lawrence C. Dodd and Bruce I. Oppenheimer (eds.), *Congress Reconsidered* (Washington, D.C.: Congressional Quarterly Press, 1981), pp. 13–30; and Norman J. Ornstein and David W. Rohde, "Shifting Forces, Changing Rules, and Political Outcomes: The Impact of Congressional Change on Four House Committees," in Robert L. Peabody and Nelson W. Polsby (eds.), *New Perspectives on the House of Representatives,* 3rd edition (Chicago: Rand McNally, 1977), pp. 186–269.

42. Lance Morrow, "A Cry for Leadership," *Time* (August 6, 1979), p. 25. For similar sentiments, see "A More Independent-Minded Institution: House, Senate Chiefs Attempt To Lead a Changed Congress," *Congressional Quarterly* (September 13, 1980), pp. 2695–2700.

43. Rayburn died November 16, 1961. During the 89th Congress, which lasted from January 1965 to December 1967, the Speaker was John W. McCormack of Massachusetts.

44. As Table 3.3 shows:

Table 3.3   Presidential strength in congressional districts, 1960–76

Year	Number of districts carried by President [a]	President's vote compared with vote for his party's successful house candidates		
		President ran ahead	President ran behind	
1960	204 (Kennedy)	22	243	(Democrats)
1964	375 (Johnson)	134[b]	158[b]	(Democrats)
1972	377 (Nixon)	104	88	(Republicans)
1976	220 (Carter)	22	270	(Democrats)

[a] Refers to the winning presidential candidate in each election.
[b] Does not include two districts where the percentage of the total district vote won by House members equaled the percentage of the total district vote won by the President.

Sources: Compiled from information in the *Congressional Quarterly Weekly Report,* Vol. 36 (April 22, 1978), p. 972; and *1967 Congressional Quarterly Almanac.*

This table is adapted from John F. Bibby, Thomas E. Mann, Norman J. Ornstein, *Vital Statistics on Congress, 1980* (Washington, D.C.: American Enterprise Institute, 1980), p. 20.

See also Hedrick Smith, "Congress and Carter: An Uneasy Adjustment," *New York Times* (February 18, 1977); David S. Broder, "Kirbo Expects Carter To Have 'Problem' on Hill," *Washington Post* (February 22, 1977).

45. Ben W. Heineman, Jr., and Curtis A. Hessler, *Memorandum for the President* (New York: Random House, 1980), p. xix.

46. See Eric L. Davis, "Legislative Reform and the Decline of Presidential Influence on Capitol Hill," *British Journal of Political Science* 9 (October 1979), pp. 465–79.

47. On access to Carter, see Eric L. Davis, "Legislative Liaison in the Carter Administration," *Political Science Quarterly* 95 (Summer 1979), pp. 287–301. See also Spencer Rich, "Shakedown Cruise: Carter's Hill Liaison Had Rough Sailing But Frank Moore Sees Smoother Seas," *Washington Post* (February 25, 1977).

48. Davis, "Legislative Liaison," pp. 288–89.

49. As Haynes Johnson recounts: "One day in December 1976, in Plains, Tip O'Neill told Carter: 'Mr. President, I want you to understand something. Some of the brightest men in America are in this Congress of the United States. Don't make the mistake of underestimating them. They've been there for years, and on any specific piece of legislation they know why every comma, every semicolon, every period is there. We want to work together, but I have a feeling you are underestimating the feeling of Congress and you could have some trouble.' Carter instantly replied: 'I'll handle them just as I handled the Georgia legislature. Whenever I had problems with the Georgia legislature I took the problems to the people of Georgia'" (*In the Absence of Power* (New York: Viking, 1980), p. 22).

50. Steven V. Roberts, "Carter Discord with Congress: President Is Apparently Seeking To Ease Strains," *New York Times* (June 5, 1979; See also Edward Walsh, " 'The Buck Stops Here': Carter Relishes Truman's Slogan," *Washington Post* (April 25, 1977); David S. Broder, "Why Carter Is Hanging Tough," *Washington Post* (April 2, 1977); Lou Cannon, "Shakedown Cruise for Carter, Hill: The Independent Democrats," *Washington Post* (May 23, 1977).

51. Elizabeth Drew, "Settling In," p. 86.

52. *Ibid.,* p. 87.

53. James T. Wooten, "Carter Gains in Confidence—And Gets a Few Lessons in the Limits of Presidential Power," *New York Times* (July 23, 1977).

54. See Charles O. Jones, "Congress and the Presidency," in Thomas

E. Mann and Norman J. Ornstein (eds.), *The New Congress* (Washington, D.C.: American Enterprise Institute, 1981), pp. 223–49 (esp. pp. 229–37).

55. See John F. Manley, "The Conservative Coalition in Congress," *American Behavioral Scientist* 17 (November-December 1973), pp. 223–47; James T. Patterson, *Congressional Conservatism and the New Deal* (Lexington: University of Kentucky Press, 1967).

56. See Jasper B. Shannon, "Presidential Politics in the South: 1938," *The Journal of Politics* 1 (May and August 1939), pp. 146–70, 278–300.

57. See Richard Bolling, *House Out of Order* (New York: E. P. Dutton, 1965), pp. 58–61; Neil MacNeil, *Forge of Democracy* (New York: McKay, 1963); and Mark F. Ferber, "The Formation of the Democratic Study Group," in Nelson W. Polsby (ed.), *Congressional Behavior* (New York: Random House, 1971), pp. 249–69.

58. Clem Miller, *Member of the House*, edited, with additional text by John W. Baker (New York: Scribner, 1962), pp. 123–24.

59. On the growth of the Democratic Study Group and its effectiveness, see Kenneth Kofmehl, "Institutionalization of a Voting Bloc," *Western Political Quarterly* 17 (June 1964), pp. 256–72; and Roger H. Davidson, "Subcommittee Government: New Channels for Policy Making," in Thomas E. Mann and Norman J. Ornstein (eds.), *The New Congress*, pp. 99–133, esp. pp. 107, 128–29.

60. Wolfinger and Arsenau show that it was mostly done through the filling of vacancies rather than through contested elections. Raymond E. Wolfinger and Robert B. Arsenau, "Partisan Change in the South, 1952–1976," in Louis Maisel and Joseph Cooper (eds.), *Sage Electoral Studies Yearbook*, vol. 4: *Political Parties: Development and Decay* (Beverly Hills: Sage Publications, 1978), pp. 179–210. See also Lawrence C. Dodd and Bruce I. Oppenheimer, "The House in Transition: Change and Consolidation," in Dodd and Oppenheimer (eds.), *Congress Reconsidered* (Washington, D.C.: Congressional Quarterly Press, 1981), pp. 31–61.

61. Ornstein and Rohde, "Shifting Forces, Changing Rules, and Political Outcomes."

62. On the revolt against Cannon, see Blair Bolles, *Tyrant from Illinois: Uncle Joe Cannon's Experiment with Personal Power* (New York: Norton, 1951); and Charles O. Jones, "Joseph G. Cannon and Howard W. Smith: An Essay on the Limits of Leadership in the House of Representatives," *The Journal of Politics* 30 (August 1968), pp. 617–46.

63. For example: Article I, Section 1, places all legislative powers in

the hands of Congress; Article I, Sections 2 and 3, gives the right of impeachment to the House of Representatives and allows the Senate to try all impeachments; Article I, Section 7, allows Congress to override a presidential veto of a bill; Article I, Section 8, among other powers, gives Congress the right to set and collect taxes and other revenues; and Article II, Section 3, allows the President to recommend legislation to Congress for its consideration.

64. Adam Clymer, "Carter's Standing Drops to New Low in Times-CBS Poll," *New York Times* (June 10, 1979); Jurek Martin, "Carter's Precarious Path to the Summit," *Financial Times* (London) (June 15, 1979), p. 22; Barry Sussman, "Poll: Carter Holding Strength," *Washington Post* (June 17, 1979); Hedrick Smith, "Carter Rating Falls in Gas Crisis; Intimates Fearful on Re-election," *New York Times* (June 29, 1979); Adam Clymer, "Gas Lack Helps Drop Carter to 26% in Poll," *New York Times* (July 13, 1979).

65. By August 1979 seven independent committees, in as many states, had registered with the Federal Election Commission to promote the Kennedy presidential candidacy. Kennedy formally disavowed them all ("Money and the Non-Candidate," *National Journal* (August 4, 1979), p. 1286). In Gallup poll trial heats Kennedy was running far better than Carter against selected Republican candidates. See the summary of results released by the Gallup poll on August 20, 1979. See also Timothy D. Schellhardt, "Criticizing Carter: Many Interest Groups Believe the President Has Let Them Down; Draft-Kennedy Move Gains," *Wall Street Journal* (June 28, 1979); Joseph F. Sullivan, "Jersey Democratic Chiefs Voicing Doubts on Cater," *New York Times* (May 29, 1979); Bill Peterson and Edward Walsh, "O'Neill Fuels Speculation on Kennedy," *Washington Post* (September 11, 1979); Adam Clymer, "Move Grows at Capitol To Urge Carter To Shun Race," *New York Times* (September 13, 1979); T. R. Reid, "Kennedy," in Richard Harwood (ed.), *The Pursuit of the Presidency* (New York: Berkley, 1981), pp. 65–66.

66. Clymer, "Carter's Standing Drops."

67. The Gallup poll summary released on August 20, 1979, showed Carter beating Reagan or Ford all through 1978, but starting at the end of June 1979 and from then onward, Ford or Reagan beat Carter.

68. David S. Broder, "Carter Seeking Oratory To Move an Entire Nation," *Washington Post* (July 14, 1979), p. A1.

69. Edward Walsh, "President Seeks Advice of More Private Citizens," *Washington Post* (July 14, 1979), p. A8.

70. Martin Schram, "Carter: Back on the Track and Eager To Retake the Lead," *Washington Post* (July 17, 1979).

71. The text of President Carter's speech is printed verbatim in the *New York Times* (July 17, 1979), p. A15, and *Congressional Quarterly Weekly Report* 37 (July 21, 1979), pp. 1470–72.

72. The texts of these speeches are given in *Congressional Quarterly Weekly Report* 37 (July 21, 1979), pp. 1472–79. On Monday, July 16, one day after the President's televised speech, a New York Times–CBS News telephone national sample survey disclosed a jump of eleven points in the number of respondents approving of Mr. Carter's "handling of his job." See Adam Clymer, "Speech Lifts Carter Rating to 37%; Public Agrees on Confidence Crisis," *New York Times* (July 18, 1979). Among those who listened to Mr. Carter's speech, 49% said they had "greater faith in President Carter's leadership" and 45% said "no." Among respondents who had had not heard the speech, only 26% said "yes" and 60% said "no."

73. Schram, "Back on the Track," July 17, 1979.

74. The most detailed and coherent recounting of these events was Elizabeth Drew, "Phase: In Search of a Definition," *The New Yorker* (August 27, 1979), pp. 45–73. I rely also upon contemporary accounts in the *New York Times, Los Angeles Times, Washington Post, Wall Street Journal, Time, Newsweek, National Journal,* and *Congressional Quarterly Weekly Report.* See in addition Joseph A. Califano, *Governing America* (New York: Simon and Schuster, 1981), pp. 425–48.

75. Terence Smith, "President Replaces Three Cabinet Secretaries: Bell, Blumenthal and Califano," *New York Times* (July 20, 1979); See also "The White House Report Card," *San Francisco Chronicle* (July 20, 1979).

76. Originally it was intended that all officials making more than $25,000 a year were to be rated until somebody pointed out to Jordan that in the Defense Department alone that involved several hundred people. Jordan retreated to requiring filled-out forms on assistant secretaries and above.

77. See Martin Schram, "Today Was the Day Hamilton Jordan Took Charge," *Washington Post* (July 18, 1979).

78. See Hedrick Smith, "Jackson Decries Cabinet Upheaval and Says Kennedy Will Run in '80," *New York Times* (July 25, 1979); Frank Lynn, "Top Democrats Disturbed by Resignations of Cabinet," *New*

York Times (July 20, 1979); "Resignation Reaction in Congress: 'Rebirth,' 'Nuts'," *Los Angeles Times* (July 18, 1979); "Cabinet Reorganization Drew Mixed Reviews," *President Carter 1979* (Washington, D.C.: Congressional Quarterly Press, 1980), pp. 5–9.

79. Steven V. Roberts, "Capital Turmoil Saddens and Confuses U.S. Aides," *New York Times* (July 21, 1979).

80. Robert Shogan, "Carter Reevaluating Hundreds of Officials," *Los Angeles Times* (July 19, 1979).

81. Mary Russell, "On the Hill One Step Forward for Carter—and One Back," *Washington Post* (July 20, 1979).

82. Jack Nelson, "Carter Shuffles Staff: Georgia Circle Intact," *Los Angeles Times* (August 11, 1979), p. 2. One congressional subcommittee chairman said of Carter's attacks on Congress: "He's trying to save himself, and it could help him, but it won't help him with his programs up here," (Steven V. Roberts, "Carter and the Congress: Doubt and Distrust Prevail," *New York Times* (August 5, 1979), p. E2).

83. James M. Perry and Albert R. Hunt, "Capital Confusion: Carter's Shakeup Stirs Tension and Disarray—Just as He Tries to Lead," *Wall Street Journal* (July 20, 1979).

84. Roberts, "Carter and the Congress" (August 5, 1979).

85. Robert D. Hershey, Jr., "Press Abroad Is Critical of Cabinet Shuffle in U.S.," *New York Times* (July 23, 1979); Jim Hoagland, "Nothing Has Changed, Carter Aides Tell Foreign Reporters," *Washington Post* (July 26, 1979).

There was a great deal of unfavorable foreign press reaction printed in the main American newspapers. A sample: "There is no longer any doubt about it—Carter has stopped governing and has started the election campaign" (a West German paper quoted in Clayton Fritchey, "A Kennedy Surrogate," *Washington Post,* July 30, 1979).

The foreign policy expert for the West German opposition party CDU, Alois Mertes, said trust in America has declined because it "cannot be decided or come about by praying. It rather has to grow from inner conviction and must root in credibility," (Michael Getler, "West Germany Says Carter's Moves Threaten NATO," *Washington Post,* July 27, 1979).

Alberto Rouchey, writing in *Corriere della Sera,* said: "The indecisiveness of the 39th President, together with his inability to hide his lack of decision has become the subject of accurate and merciless analysis"; *Die Zeit* criticized the firings as incomprehensible. "Incomprehensible because, after all, he hired those he is now firing. He has poorly coordinated and inadequately led them. And isn't he himself the very 'Washington' he now criticizes?" (Both in Robert D. Hershey, Jr., "Press Abroad Is Critical of Cabinet Shuffle in U.S." *New York Times,* July 23, 1979.)

Time said: ". . . Carter's handling of the mass firings caused Europeans to cluck in wonder. A high ranking West German Foreign Ministry official asked: 'Is this serious, or is this just a great religious exercise for the soul?' "

"Oslo's middle-roading daily *Verdens Gang* called the Washington situation a 'circus' and a 'balancing act without a safety net.' "

"Concluded London's conservative *Daily Mail:* 'to revive his personal standing with voters before the next presidential election looks more like a narcissistic charade than a national crusade.' "

"*The Economist* suggested that Carter's efforts would not be successful, 'unless there is some understanding of how the world works and some readiness to eschew symbolism and appearances and go instead after the substance of the problems.' "

"One money trader in Milan summed it up as follows: 'When Carter speaks, the dollar plummets' " ("Slumping to a New Low Abroad," *Time* (July 30, 1979), pp. 23–24).
86. Michael Getler, "Carter's Shakeup Worries West Germans," *Washington Post* (July 21, 1979); Hobart Rowen, "Resignation Offer Shocking Abroad," *Washington Post* (July 19, 1979).
87. Republican reactions included the following:

"If somebody told me that Anastazio Somoza had just been named to replace Harold Brown as Secretary of Defense," Rep. Guy Vander Jagt (R-Mich.) said to a colleague on the House floor in the early afternoon, "I'd run out to see if it was on the news ticker in the lobby" (David S. Broder and Ward Sinclair, "A Political Skylab Hits the Capital," *Washington Post,* July 20, 1979).

Senator Lowell Weicker: "I think the President is nuts. This is just a continuation of the scapegoat operation. It is the President, not his Cabinet, that the American people have lost confidence in" ("Resignation Reaction in Congress: 'Rebirth,' 'Nuts'," *Los Angeles Times,* July 18, 1979).

Senate Republican Whip Ted Stevens, said: "Some of us are seriously worried that he might be approaching some sort of mental problem. He ought to take a rest" ("Carter's Great Purge," *Time* (July 30, 1979), pp. 10–16).
88. Democratic responses included the following:

Senator Abraham Ribicoff of Connecticut "described the mass resignation as 'a very foolish move.' He added: 'There's no justification for blanket resignations. The President should know how good a member of his Cabinet or staff is' " (Frank Lynn, "Top Democrats Disturbed by Resignation of Cabinet," *New York Times,* July 20, 1979).

House Majority Whip John Brademas: "He's like a football player who's behind in the game, then catches the ball and is breaking through to daylight, when he suddenly runs out of bounds" (Mary Russell, "On the Hill: One Step Forward for Carter—and One Back," *Washington Post,* July 20, 1979). "Rep. William Ford (D-Mich.), chairman of an Education and Labor subcommittee, said the actions were 'disastrous' for a higher education bill he [Carter] is working on" (*ibid.*).

"Said Tim Hagen, the Cleveland area Democratic Party chairman: 'In baseball, you fire the manager. Here they are asking the players to quit.' Sniped the Massachusetts Democratic Party chairman, Chester Atkins: 'The mouse that roared is still a rodent.' " "Said Democratic Congressman Charles Wilson of Texas: 'Good grief! They're cutting down the biggest trees and keeping the monkeys' " ("Carter's Great Purge," *Time* (July 30, 1979), pp. 10–16).

"David Obey of Wisconsin called the dismissals 'a victory for mediocrity—a dunderheaded performance' and warned that they would weaken President Carter's chances of renomination" (Hedrick Smith, "Dismissals Taken as Pre-Campaign Move by Carter," *New York Times,* July 20, 1979).

89. "The Cabinetmaker Selects His New Timber," *The Economist* (July 28, 1979), pp. 39–43.

90. See Tom Morgenthau with James Doyle, "The Mood of a Nation," *Newsweek* (August 6, 1979), pp. 26–27.

91. Caddell's data were of the following sort: In a national opinion survey, people were asked to rate the state of the country and the state of their own lives five years ago, today, and five years from now, on a scale of one to ten. Because what "the state of" the country and one's own life are supposed to mean is intentionally left unspecified, so as to evoke whatever is salient to respondents, this scale is regarded as a generalized, self-administered measurement of optimism-pessimism. Caddell's data showed a decline in expectations of the future. It is unclear how responses were aggregated to arrive at this conclusion or what inferences might sensibly be drawn from it. Needless to say, Caddell's analysis was not widely shared among public opinion analysts, as *Newsweek* discovered: "Says one opinion expert, the University of Michigan's F.

Thomas Juster: 'My instinct is that it's an exaggerated story. It just doesn't make any sense' "(p. 26).

"Political scientist Warren Miller also at the University of Michigan, says the degree of voter alienation has been 'grossly overinterpreted.' 'There is no massive and personal disaffection from society—or government,' Miller says. Instead, he adds, 'there is clear dissatisfaction with the Carter Presidency' " ('The Mood of a Nation," *Newsweek* (August 6, 1979), p. 27). See also Patrick H. Caddell, "Crisis Of Confidence I: Trapped in a Downward Spiral," *Public Opinion* 2 (October/November 1979), pp. 2–7; Warren E. Miller, "Crisis of Confdence II: Misreading the Public Pulse," *ibid.*, pp. 9–15.

92. Thomas M. DeFrank, "How Carter Sees It," *Newsweek* (July 30, 1979), p. 25. See also Terence Smith, "On Carter and the Washington Press Corps," *New York Times* (August 2, 1979).

93. Terence Smith, "Reporter's Notebook: At the White House Jeans Are Passé and Neckties De Rigueur," *New York Times* (July 27, 1979).

94. See Charles Mohr, "Jordan Says Carter Will Widen His Circle of Advisers," *New York Times* (July 23, 1979).

95. See Knott and Wildavsky, "Jimmy Carter's Theory of Governing."

96. See Elihu Katz and Paul F. Lazarsfeld, *Personal Influence* (Glencoe, Ill.: Free Press, 1955).

97. The literature on opinion formation among members of Congress is by now voluminous. See John W. Kingdon, *Congressmen's Voting Decisions* (New York: Harper & Row, 1973); Aage R. Clausen, *How Congressmen Decide* (New York: St. Martin's Press, 1973); Donald R. Matthews and James A. Stimson, *Yeas and Nays: Normal Decision Making in the U.S. House of Representatives* (New York: John Wiley, 1975); Lewis Anthony Dexter, "What Do Congressmen Hear?," in Nelson W. Polsby (ed.), *Congressional Behavior* (New York: Random House, 1971), pp. 28–41; and Dexter, *The Sociology and Politics of Congress* (Chicago: Rand McNally, 1969).

98. Dom Bonafede, "White House Report: The Fallout from Camp David—Only Minor White House Changes," *National Journal* (November 10, 1979), pp. 1893–97.

99. Joseph Califano claims that in his 2½ years as secretary of H.E.W. he had spent only two hours with Jordan and that Jordan seldom returned the telephone calls of members of the cabinet (*Governing America*, p. 441).

100. " 'I'm sure I've met him,' Senate Majority Leader Robert Byrd loftily remarked last week. 'But I've never had a conversation

with him' " ("Here Comes Mr. Jordan," *Time* (July 30, 1979), p. 22).

101. James M. Perry, "Capitol Kept in Suspense as Carter Mulls Resignation Offers of Top Aides, Cabinet," *Wall Street Journal* (July 19, 1979).
102. "Bell, Departing, Is Given President's Warm Praise" *New York Times* (August 5, 1979), p. 42.
103. "Jordan's 'Beautiful' But Who's Jerdun?" *New York Times* (July 28, 1979).

One immediate addition in White House staffing was the employment on the senior staff of Hedley Donovan, the recently retired editor-in-chief of Time Inc. publications, early Carter boosters. It was not made clear what Mr. Donovan's responsibilities would be, nor did it ever become clear in the months after, but it could not escape the notice of observers that Mr. Donovan's predominant expertise and life experience were as a New York-based executive of a national news publication. Administration spokesmen insisted that Mr. Donovan's job would be substantive, connected neither with public relations nor politics. The announcement of his employment was greeted with fulsome approbation by *Time* magazine (August 6, 1979), pp. 20–21. Said *Time:* "Donovan expresses himself with conviction and candor. His sound, unruffled judgment . . . has been one of his great strengths." See also John Osborne, "Hedley the Don," *The New Republic* (January 26, 1980), pp. 9–10.

104. Terence Smith, "At the White House Jeans Are Passé," *New York Times* (July 27, 1979)
105. According to the CBS News/ *New York Times* exit poll on November 4, 1980, the Democratic defection rate was 25 percent. See Adam Clymer, "The Democrats Look for New Ideas, and Jobs," *New York Times* (November 9, 1980); and Clymer, "Displeasure with Carter Turned Many to Reagan," *ibid.* William Schneider writes: "Former Carter supporters were four times as likely as former Ford supporters to report that they "sat out" the 1980 presidential election. Abstention siphoned off even more Democratic voters than did defection to other candidates. The result was that Carter was able to hold on to only a little more than a third of those who had given him his scanty majority four years earlier, while Reagan retained two-thirds of the former Ford voters . . . . The 1980 election was essentially a referendum on the Carter administration. . . ." (Schneider, "The November 4 Vote for President: What Did It Mean?," in Ranney (ed.), *The American*

Elections of 1980) pp. 212–62.) See also Nelson W. Polsby, "Party Realignment in the 1980 Election," *Yale Review* (Fall 1982).

106. See Kenneth Reich, "State Democrats Bitter on Carter's Early Concession," *Los Angeles Times* (November 5, 1980); William Endicott, "Anti-Carter Vote Seen in State Races," *ibid.* (November 6, 1980); Roger Smith, "Early Concession Costly: State Party Leaders Sure Carter 'Paralyzed' Effort," *ibid.* (November 7, 1980); John Fogarty, "Carter Saw No Harm in Conceding Early," *San Francisco Chronicle* (November 7, 1980); and John Balzar, "Demos Hurt by Early TV Vote Report," *ibid.* (March 11, 1981); Jack W. Germond and Jules Witcover, *Blue Smoke and Mirrors* (New York: Viking, 1981), pp. 314–15.

107. The actual results in the Iowa precinct caucuses on January 19, 1976, were: Uncommitted 38.5; Carter 29.1; Bayh 11.4; Harris 9.0; Udall 5.8; Shriver 3.1; Jackson 1.1; the rest less than 1%. In the New Hampshire primary of February 24, 1976, the results were: Carter 29.4, Udall 23.9, Bayh 16.2, Harris 11.4, Shriver, 8.7, Humphrey (write-ins) 5.6, Wallace (write-ins) 1.4, Jackson (write-ins) 1.4, Ellen MacCormick, 1.3 (*Congressional Quarterly Weekly Report* (February 28, 1976), p. 459; and *ibid.* (July 10, 1976), p. 1810).

108. See Nelson W. Polsby, "The Democratic Nomination," in Ranney (ed.), *The American Elections of 1980*, pp. 37–60.

Chapter IV: Wider Consequences

1. If there is a tendency, the consensus seems to be that it is "liberal." Half of the Washington reporters in Stephen Hess's sample said they thought there was a political bias in the Washington news corps, and of these, 96 percent said the bias was liberal in its direction. Most Washington reporters identified themselves as something other than liberal, however. See Stephen Hess, *The Washington Reporters* (Washington, D.C.: Brookings Institution, 1981), pp. 87ff. Another recent survey of the political and social opinions of influential U.S. journalists places them as predominantly liberal, as favoring welfare and capitalism, as well-educated and unalienated. See S. Robert Lichter and Stanley Rothman, "Media and Business Elites," *Public Opinion* (October/November 1981), pp. 42–46, 59–60.

2. When the McGovern Commission adopted guideline A-2, designed to encourage national convention representation of "minority groups, young people and women in reasonable relation-

ship to their presence in the population . . ." it was not explicit about which groups qualified as minorities. The Commission report, *Mandate for Reform,* only documented what was regarded as disproportionate absence of black delegates (pp. 26–27). No other minority groups were discussed. The *Delegate Selection Rules for the 1980 Democratic National Convention* (Washington, D.C.: Democratic National Committee, June 9, 1978) cited the need for state parties to develop and submit Affirmative Action Programs for women, blacks, Hispanics, and Native Americans (p. 7, rule 6A). The Commission on Presidential Nomination and Party Structure (Morley Winograd, chairman) likewise referred to a "particular concern for minority groups, Native Americans, women, and youth, in the delegate selection process," in its recommended delegate selection rules (*Openness, Participation and Party Building: Reforms for a Stronger Democratic Party* (Washington, D.C.: Democratic National Committee, January 25, 1978), pp. 96–97, rule 18). And contained in the 1974 Mikulski Commission report was the following affirmative action rule: "In order to encourage full participation by all Democrats, with particular concern for minority groups, native Americans, women and youth, in the delegate selection process and in all party affairs, the national and state Democratic Parties shall adopt and implement affirmative action programs" (*Congressional Quarterly Weekly Report* (March 9, 1974), p. 607).

3. An interesting case in point is a group on the national political scene that has only recently been given much media visibility: the physically handicapped. The processes of deliberation leading up to the proclamation of sweeping guidelines concerning provision for the handicapped of access to public facilities by the Secretary of HEW, Joseph Califano, were greatly hastened by televised demonstrations and, according to Califano's account, by Califano's fears of lurid, unfavorable headlines (see *Governing America* (New York: Simon and Schuster, 1981), pp. 258–62). It may be instructive for a reader to compare the justifications given therein with those of Jacobus tenBroek in "The Right To Live in the World: The Disabled in the Law of Torts," *California Law Review* 54 (May 1966), pp. 841–919.

4. See the essay on Dunlop in *Current Biography 1951,* pp. 173–74; and the account of the appointment of Ray Marshall, who got the job in the Carter administration in *Current Biography 1977,* p. 287; also Bruce Adams and Kathryn Kavanagh-Baran, *Promise and Performance* (Lexington, Mass.: Lexington Books, 1979), pp. 38–39, 42.

5. "Some Snags in the Stretch," *Time* (December 27, 1976), p. 6:

". . . Carter aides asked AFL-CIO officials to suggest alternatives to Dunlop who would be acceptable to labor boss George Meany. Back came the word: 'His first, second, third and fourth choices are Dunlop.' " Lane Kirkland, AFL-CIO secretary-treasurer said: "He comes as close to being the indispensable man as there is" (Edward Cowan, "Labor Leader Says Economy Needs Permanent Tax Cut of $25 Billion," *New York Times,* December 14, 1976).

6. Eileen Shanahan, "17 Groups Seek a Labor Secretary Committed To Fight Discrimination," *New York Times* (December 11, 1976); Tom Wicker, "The Dunlop Signal," *New York Times* (December 14, 1976); "The Cabinet Problem," *New York Times* (December 20, 1976). See also Robert G. Kaiser, "Women Lobby for Role in Carter Camp," *Washington Post* (December 5, 1976).

7. See Adams and Kavanagh-Baran, *Promise and Performance;* see also Nancy Hicks, "Feminists Critical of Carter on Jobs," *New York Times* (February 8, 1977); Edward Walsh, "Costanza Defends Carter on Recruitment of Women," *Washington Post* (January 28, 1977); "Carter Appointments: 13% Are Women, 11% Are Black," *Washington Post* (June 19, 1977).

8. See, for example, Charles McC. Mathias, Jr., "Ethnic Groups and Foreign Policy," *Foreign Affairs* (Summer 1981), pp. 975–98, for criticism of Jewish influence in foreign affairs.

9. See Theodore H. White, *The Making of the President 1972* (New York: Atheneum, 1973), pp. 263–80; Gordon L. Weil, *The Long Shot* (New York: Norton, 1973), pp. 156–94; Timothy Crouse, *The Boys on the Bus* (New York: Random House, 1973), pp. 325–33; Gary Hart, *Right from the Start* (New York: Quadrangle, 1973), pp. 250–54, 256–65; and Douglas E. Kneeland, "Behind Eagleton's Withdrawal: A Tale Of Confusion and Division," *New York Times* (August 2, 1972).

10. In addition to several editorials, the *Washington Post* provided a summary description of press opinion. Twenty-three newspapers in their survey commented adversely and ten (including the *Post* itself) called for Eagleton's withdrawal from the ticket. Among the others seeking Eagleton's withdrawal were the *New York Times, Los Angeles Times, Baltimore Sun,* and *Newsday.* The *New York Post* was quoted as saying: "In this year of great national decision . . . [Eagleton] had disqualified himself by his apparent act of concealment . . ." (Jack Fuller, "Editorials Are Mixed on Eagleton," *Washington Post,* July 28, 1972). See also *Washington Post* editorials on July 27, 29, 31, and August 2, 1972.

11. Hart, *Right from the Start,* p. 265.

12. See Timothy Crouse, *The Boys on the Bus,* pp. 320–33. One might

compare this episode with Dwight Eisenhower's handling of the disclosure of the Nixon fund, which jeopardized Richard Nixon's place on the Republican ticket in 1952. There was considerable pressure from the newspapers—notably the *New York Herald Tribune*—on Nixon to withdraw, but Nixon's famous television speech—the Checkers speech—presumably saved him from disgrace. In addition, several political friends and allies of General Eisenhower seem to have played important roles in determining the outcome, whereas accounts of the Eagleton episode give the impression that McGovern was more or less on his own. See Earl Mazo and Stephen Hess, *Nixon: A Political Portrait* (New York: Popular Library, 1968), pp. 76–125.

13. Alexis de Tocqueville, *Democracy in America* (New York: Knopf Vintage, 1956) is a classic statement. More modern discussions include David B. Truman, *The Governmental Process* (New York: Knopf, 1971); and Robert H. Salisbury, "Interest Groups," in Fred I. Greenstein and Nelson W. Polsby (eds.), *Handbook of Political Science* (Reading, Mass.: Addison-Wesley, 1975), vol. 4, pp. 171–228.

14. Indications of changes in the average size of American nuclear families—from 4.11 in 1930 to 2.73 in 1981—have important political consequences. For a rough estimate of the decline of primary group intermediation and the rise of intermediation through the mass media, compare these numbers with those on the number of television sets per household in 1950 (.101) and 1979 (1.67), and average viewing per day in 1950 (4.6 hours) and 1980 (6.3 hours). Bureau of Census, *Statistical Abstract of U.S. 1979* (Washington, D.C., 1979).

15. There are a number of indices that show how technology has multiplied the options of individual Americans to engage in communication with one another over long distances. Note, for example, the following:

Table 4.1 Increased capabilities for long-distance interaction among Americans

	1920	1950	1978
Auto registrations (thousands)	9,239	48,567	117,100
Telephones in use (millions)	13	43	169
Domestic airways passenger miles (billions)		8	182.7

Source: *Statistical Abstract 1979*, pp. 581–82, 642, 667.

16. See Nelson W. Polsby, "Legislatures," in Greenstein and Polsby (eds.), *Handbook of Political Science,* vol. 5, pp. 257–319.

17. Among the writers interested in the ways in which social incentives shape organizational and political behavior and who have dealt in detail with problems similar to this one I particularly recommend James Q. Wilson. See his *The Amateur Democrat* (Chicago: University of Chicago Press, 1966) and *Political Organizations* (New York: Basic Books, 1973), esp. chapter 3.

18. Examples of such interests in American political history abound: abolitionism and the temperance movement are two of the most famous. Among the classic studies of single issue interest groups in the twentieth century are E. E. Schattschneider, *Politics, Pressures and the Tariff* (New York: Prentice-Hall, 1935); Peter H. Odegard, *Pressure Politics: The Story of the Anti-Saloon League* (New York: Columbia University Press, 1928); Clement E. Vose, *Constitutional Change* (Lexington, Mass.: Lexington Books, 1972), esp. pp. 69–137; and Raymond Bauer, Ithiel de Sola Pool, and Lewis Anthony Dexter, *American Business and Public Policy* (New York: Atherton, 1963).

19. The best single summary I have seen on the turnout question is Richard Brody, "The Puzzle of Political Participation in America," in Anthony King (ed.), *The New American Political System* (Washington, D.C.: American Enterprise Institute, 1978), pp. 287–324. See also William J. Crotty and Gary C. Jacobson, *American Parties in Decline* (Boston: Little, Brown 1980).

20. A particularly vigorous argument to this effect is made by Byron Shafer in his unpublished paper "Reform and Alienation: The Decline of Intermediation in the Politics of Presidential Selection" (mimeo, Russell Sage Foundation, 1981).

21. For example consider some of the seemingly disparate findings in Jack Dennis's important article "Trends in Public Support for the American Political System," *British Journal of Political Science* 5 (April 1975), pp. 187–230. Dennis reports that the numbers of people were on the increase who said:

(1) Parties "make a difference" and are ideologically distinguishable
(2) Divided government is preferable
(3) Parties were less interested in what respondents thought, less controllable by citizens
(4) They had personally campaigned or donated money to politics.

These findings—and especially when taken together—are compatible with distaste for and worry about political extremism. So are the trend lines on many of Dennis's indicators of public support that break downward with the 1964 "choice not an echo" presidential election (see his figure 2.). See also Norman H. Nie, Sidney Verba, and John R. Petrocik, *The Changing American Voter* (Cambridge, Mass.: Harvard University Press, 1976).

22. A good summary is Jacob Citrin's "The Alienated Voter," *Taxing and Spending* (October–November 1978), pp. 7–11. See also Arthur H. Miller, "Political Issues and Trust in Government," *American Political Science Review* 68 (September 1974), pp. 951–72; and Jack Citrin, "Comment: The Political Relevance of Trust in Government," *ibid.*, pp. 973–88.

23. This is true, for example, of the propensity of voters to turn out and vote, but not true, as Dennis ("Trends in Public Support") shows, of the willingness to give money or to campaign. See Richard Brody, "The Puzzle of Political Participation."

24. See Samuel P. Huntington, "The Democratic Distemper," *The Public Interest* 41 (Fall 1975), pp. 9–38; Aaron Wildavsky, "The Past and Future Presidency," *ibid.*, pp. 56–76; and Wildavsky, "Government and the People," *Commentary* 56 (August 1973), pp. 25–32.

25. The Hughes Commission wrote: "A convention of delegates selected by party officials is open to the same criticism as direct appointment of delegates to the National Convention—party officials do not necessarily represent the party membership on issues of presidential politics. The Commission's study indicates that over 600 delegates to the 1968 Convention were selected by processes which have included no means of voter participation since 1966" (Commission on the Democratic Selection of Presidential Nominees, *The Democratic Choice* (Washington, D.C., 1968), p. 24). See also William J. Crotty, *Decision for the Democrats* (Baltimore: Johns Hopkins University Press, 1978), pp. 254–57.

26. The locus classicus is Bernard C. Cohen, *The Press and Foreign Policy* (Princeton: Princeton University Press, 1963). Further explorations of themes first identified by Cohen can be found in a series of Harvard doctoral dissertations, e.g., Edward Jay Epstein, *News from Nowhere* (New York: Random House, 1973); Leon Sigal, *Reporters and Officials* (Lexington, Mass: D. C. Heath, 1973); and Paul Weaver, "The Metropolitan Newspaper as a Political Institution," P.h.D. dissertation, 1969; Paul Weaver, "How the *Times* Is Slanted Down the Middle," *New York* (July 1, 1968), pp. 32–36. See also Herbert J. Gans, *Deciding What's News* (New York: Vintage Books, 1980).

27. The phrase "adversary culture" is Lionel Trilling's and comes from his discussion of modernism in literature in *Beyond Culture,* (New York: Viking, 1965), pp. xii ff. It was pressed into service in the present context·by Daniel Patrick Moynihan in "The Presidency and the Press," *Commentary* 51 (March 1971), p. 43 (reprinted in Moynihan, *Coping* (New York: Random House, 1973), pp. 318–20). See also Joseph Kraft, "The Imperial Media," *Commentary* (May 1981), pp. 36–47; Max M. Kampelman, "The Power of the Press: A Problem for Our Democracy," *Policy Review* 6 (Fall 1978), pp. 7–39; and Michael Schudson, *Discovering the News* (New York: Basic Books, 1978), pp. 176ff. The essental text on "radical chic" is Tom Wolfe, "Radical Chic," pp. 3–94 of *Radical Chic and Mau-Mauing the Flak Catchers* (New York: Farrar, Straus, and Giroux, 1970).

28. Lichter and Rothman, *op. cit.,* give evidence of liberalism but not of much radicalism in the attitudes of their influential news media respondents.

29. See Paul H. Weaver, "Captives of Melodrama" *New York Times Magazine* (August 29, 1976), pp. 6, 48, 50–51, 54, 56–57; Thomas E. Patterson, *The Mass Media Election: How Americans Choose Their President* (New York: Praeger, 1980).

30. Robert G. Kaiser described some of these in the course of his reports on the television coverage of the 1980 election for the *Washington Post.* His conclusion on October 10: "There are three stock formats for the basic television news report on the 1980 campaign. Format One might be called 'Charge-Countercharge' . . . Format Two is . . . a day in the candidate's life on the road . . ·. Format Three is the handicapper's report from one of the major racetracks, an evaluation of the horse race in one of the big states" ("T.V. on the Trail: A Three-Course Menu for Fluff"). Similar comments have been made about print journalism. See Paul Weaver, "The Politics of a News Story," in Harry M. Clor (ed.), *The Mass Media and Modern Democracy* (Chicago: Rand McNally, 1974), pp. 85–112. Weaver's view of the components of the typical news story is summarized by Schudson (p. 185), as follows:

> . . . The bias is toward statements of fact which are observable and un-ambiguous; toward broad, categorical vocabulary—"say" rather than "shout" or "insist"; toward impersonal narrative style and "inverted pyramid" organization which force a presentation of facts with "as little evocation of their real-world context" as possible; toward conflicts rather than less dramatic happenings; toward "events" rather than processes.

Cohen, *The Press and Foreign Policy,* comes to similar conclusions.

31. The *Post*'s ombudsman in 1981, Bill Green, wrote: "There is no question about the pressures and competition in the *Washington*

Post's newsroom. They are powerful. Some people flourish, others get crushed. It is major-league journalism. 'Hardball' as [*Post* Executive Editor] Ben Bradlee describes it. The troubling question is whether pressure on the staff distorts the news published in the paper" ("The Pressures: Heat and the Achievers Both Have a Tendency to Rise," *Washington Post,* April 19, 1981).

32. See Daniel Machalba, "UPI Struggles as It Loses Ground to AP, Other News Services," *Wall Street Journal* (July 11, 1979).

33. "Reporters Appraise Campaign Roles," *New York Times* (February 1, 1982).

34. The belief by news organizations that impact on political actors is important constitutes at least one explanation for the odd spectacle of the *Los Angeles Times* taking out full-page ads in the *New York Times* and *Washington Post* in order to display notable instances of its political coverage to audiences it does not ordinarily have. See also David Halberstam, *The Powers That Be* (New York: Dell, 1979), pp. 548–49, 1006.

35. Here is an unself-consciously comic example of the process at work: "The United States finds it hard to understand why King Hussein . . . joined the outcry against Camp David. . . . But there are reasons for the policy from the Jordanian viewpoint. They were explained by me by Hussein's articulate brother, Crown Prince Hassan— the King is abroad—and by high officials" (Anthony Lewis, "At Home Abroad: When Friends Fall Out," *New York Times* (April 16, 1979), p. A17).

36. The ground rules—"Lindley rules"—which define the mutual expectations of reporters and news sources under "off the record," "deep background," "background," and "on the record" circumstances were codified by Ernest K. Lindley of *Newsweek.* See William J. Small, *Political Power and the Press* (New York: Norton, 1972), pp. 178–79. A *New York Times* reporter, commenting on quotations embarrassing to Budget Director David Stockman in the August 1981 *Atlantic Monthly* said: "In Washington, editors frequently meet socially with government officials and other people in the news to talk informally about public affairs. Such meetings do not generally produce news articles directly, but may encourage an editor to assign a reporter to pursue on the record what an editor has been told off the record" (Jonathan Friendly, "Post Allowed Editor to Scoop Newspaper," *New York Times* (November 13, 1981), p. 39). See, in general, Lou Cannon, *Reporting: An Inside View* (Sacramento: California Journal Press, 1977); Joseph Alsop and Stewart Alsop, *The Reporter's Trade* (New York: Reynal, 1958); William L. Rivers, *The Opinionmakers* (Boston: Beacon Press, 1965),

especially pp. 34–39; and Robert Pierpoint, *At the White House* (New York: Putnam, 1981).

37. In the field of national news, the most important general news organizations are the three television networks, the two wire services, the two big news magazines, and the handful of big newspapers owning their own national syndication services, principally the *New York Times* and the *Washington Post–Los Angeles Times*. For evidence of heavy dependence on wire service coverage of national and political news by local newspapers scattered all across the United States, see Michael J. Robinson and Margaret H. Sheehan, *Over the Wire and on TV* (New York: Russell Sage, 1982), chapter 2.

     Wire service reports frequently dominate the production of news in the first place, according to the observations by Timothy Crouse of political campaign reporting in 1972. See *The Boys on the Bus* (New York: Random House, 1972), p. 22: "Most reporters . . . followed the wire service men whenever possible. Nobody made a secret of running with the wires; it was an accepted practice."

     Readers will recognize that differentiated marketing strategies, aiming mostly at businessmen, are at work for otherwise important news organizations like the *Wall Street Journal* and *U.S. News and World Report,* for whom deductions from the axiom of competitiveness must be suitably modified. Stephen Hess, *The Washington Reporters* (Washington, D.C.: Brookings Institution, 1981), p. 140, gives a list of "influential" news media similar to the one above, based on his interviews with Washington reporters, and their reading and listening habits, subtracting the *Los Angeles Times* and including the now-defunct *Washington Star,* the *Wall Street Journal,* and *U.S. News and World Report.*

38. See Bernard C. Cohen, *The Press and Foreign Policy,* pp. 58ff.

39. A theory of the underlying process of social definition and its consequences was worked out years ago in some detail by W. I. Thomas. See Edmund H. Volkart (ed.), *Social Behavior and Personality: Contributions of W. I. Thomas to Theory and Social Research* (New York: Social Science Research Council, 1951).

40. Vice President Agnew's famous contribution to media criticism ("nattering nabobs of negativity" and so on) was a single speech given on November 13, 1969, in Des Moines, Iowa, before a meeting of the Midwest Republican Conference. It was written by Pat Buchanan and had President Nixon's prior approval. See Evans and Novak, *Nixon in the White House: The Frustration of Power* (New York: Random House, 1971), pp. 315–17. Other, more measured complaints about news media treatment of President Nixon be-

fore Watergate are contained in James Keogh, *President Nixon and the Press* (New York: Funk and Wagnalls, 1972).

41. A complementary view from another part of the ideological spectrum, that concentration of the ownership of newspapers in unworthy hands leads to inaccurate or malicious news coverage, seems to me about equally limited in its explanatory power, especially as far as national news coverage is concerned. It cannot be ruled out in some circumstances—e.g., for many decades in Los Angeles (see David Halberstam, *The Powers That Be,.* p. 167), in Wilmington, Delaware (see Ben H. Bagdikian, *The Effete Conspiracy*, New York: Harper & Row, 1972, pp. 69–79), or Manchester, New Hampshire (see Eric Veblen, *The Manchester Union Leader in New Hampshire Elections*, Hanover, New Hampshire: University Press of New England, 1975)—but will not cover as many cases or constraints upon outcomes as the normal exercise of professionalism under competitive conditions such as I am describing.

42. See Nelson W. Polsby, "Toward an Explanation of McCarthyism," *Political Studies* 8 (October 1960), pp. 250–71.

43. As a fine recent book on McCarthy's press coverage documents. See Edwin R. Bayley, *Joe McCarthy and the Press* (Madison: University of Wisconsin Press, 1981).

44. "Wirthlin is a regular visitor at the White House. In one two-week period early this month, he said, he had two private meetings with the president. Sometimes he meets with Mr. and Mrs. Reagan together. He also has private sessions with the senior presidential aides, and with the secretary of state, Alexander M. Haig, Jr." (Robert G. Kaiser, "White House Pulse-Taking," *Washington Post*, February 24, 1982). See also B. Drummond Ayres, Jr., "G.O.P Keeps Tabs on Nation's Mood," *New York Times* (November 16, 1981); and Dom Bonafede, "As Pollster to the President, Wirthlin Is Where the Action Is," *National Journal* (December 12, 1981), pp. 2184–88.

45. See Wendell Rawls, Jr., "Carter's Poll-Taker Seems To Voice 1980 Catchwords," *New York Times* (August 14, 1979); E. J. Dionne, Jr., "Polls, Once Scorned, Gain New Esteem," *New York Times* (April 5, 1980); E. J. Dionne, Jr., "The Business of the Pollsters," *New York Times* (June 29, 1980).

46. Patricia Meisol, "Searching for a Cure for the Proposition 13 Epidemic," *National Journal* (August 26, 1978), pp. 1362–65. ". . . voters in eight states and some counties will be confronted with ballot proposals on Nov. 7 that would limit taxing or spending or both. Three of them—in Idaho, Michigan and Oregon—are lifted almost verbatim from the constitutional amendment that Califor-

nia voters approved overwhelmingly on June 6" (p. 1362). The others were in Arizona, Colorado, Nebraska, South Dakota, and Texas.

47. As James S. Coleman's work has documented. See *Community Conflict* (Glencoe, Ill.: Free Press, 1957), p. 19.

48. This follows the usage of Frederick M. Watkins. See Watkins, *The Age of Ideology: Political Thought, 1750 to the Present* (Englewood Cliffs, N.J.: Prentice-Hall, 1965).

49. See, for example, Edward C. Banfield, *Political Influence*, (New York: Free Press of Glencoe, 1961).

50. See, for example, David Sanford, *Me and Ralph* (Washington: New Republic Book Co., 1976).

51. *Congressional Quarterly* keeps annual account of interest group ratings of Congressmen according to their recorded votes on issues of interest to them. Americans for Democratic Action and the AFL-CIO Committee on Political Education maintain, roughly speaking, left-wing score-cards; Americans for Constitutional Action is, more or less, right-wing.

Table 4.2   Voting records of Senator Jackson and Representative Udall

Year	ADA		COPE		ACA	
	Jackson	Udall	Jackson	Udall	Jackson	Udall
1961–2	100	100	100	80	3	18
1963–4	82	88	73	82	6	7
1965–6	80	82	100	0	11	8
1967	69	93	100	100	0	4
1968	57	92	100	100	12	0
1969	78	67	100	80	14	0
1970	56	76	100	83	24	0
1971	56	81	100	82	27	4
1972	40	100	100	100	38	0
1973	55	84	100	82	21	8
1974	62	65	82	100	11	8
1975	61	47	90	92	8	13

Each score is the percentage of selected votes on which Jackson or Udall agreed with the given organization's position. The appropriate votes and scores are calculated by the interest groups and compiled by *Congressional Quarterly*. A high score indicates agreement between Udall or Jackson and the organization. A traditional liberal Democrat would be expected to receive high ADA and COPE scores and low marks on the ACA scale. More specific information on the scoring is in *Congressional Quarterly Weekly Report* (March 21, 1981), p. 516.

Source: *Congressional Quarterly.*

52. See Norman J. Ornstein, "The House and the Senate in a New Congress," in Thomas E. Mann and Ornstein (eds.), *The New Congress* (Washington, D.C.: American Enterprise Institute, 1981), pp. 363–83. Also Norman J. Ornstein (ed.), *Congress in Change* (New York: Praeger, 1975).

53. For a general discussion of tensions between political executives and the permanent government, see Hugh Heclo, *A Government of Strangers* (Washington, D.C.: Brookings Institution, 1977).

54. Data on resignations of senior officials is highly suggestive. "About 95 percent of top bureaucrats reaching retirement age—those between 55 and 59 with 30 years of federal experience—are deciding to leave, compared with about 18 percent in 1978, the General Accounting Office (GAO) reports" (p. 1296). William J. Lanouette, "SES: From Civil Service Showpiece to Incipient Failure in Two Years," *National Journal* (July 18, 1981), pp. 1296–99. This trend may be variously attributed to the deleterious effects of the pay cap on senior civil servants coupled with an uncommonly generous pension program—both conditions which obtained in only slightly diminished degree in 1978—or to the governmental reforms of the Carter administration, which effectively deprived senior civil servants of tenure in office.

55. Jeffrey M. Berry, *Lobbying for the People: The Political Behavior of Public Interest Groups* (Princeton: Princeton University Press, 1977), p. 186.

56. *Ibid.*, p. 188.

57. To mention only some of the largest: The Sierra Club has approximately 270,000 members nationally, Common Cause 225,000, Environmental Defense Fund 47,000, Natural Resources Defense Council 40,000, and Friends of the Earth 25,000.

58. William Cavala, "Changing the Rules Changes the Game: Party Reform and the 1972 California Delegation to the Democratic National Convention," *American Political Science Review* 68 (March 1974), pp. 27–42, is an early discussion of some of these issues in the context of delegate selection politics. On the difficulties of nominal versus authorized membership in a politically significant group, see, for example, the unpublished "Testimony of Lilia Molina Before the National Democratic Committee's Commission on Presidential Nominations" (mimeographed, Democratic National Committee, October 16, 1981). Ms. Molina was concerned about Hispanic members of the California delegation to the 1980 Democratic National Convention, and, among other observations, pointed out: "(1) Jim Costa, listed as Hispanic, is of Portuguese

ancestry, and even though technically Hispanic, does not identify
with the majority Hispanic group in California: Chicanos; (2) Louis
Papan, listed as Hispanic, is of Greek Ancestry; and (3) Houston
B. Quick, listed as Hispanic, was not listed as such by the California Democratic Party" (p. 2).
59. By attacking Congressional use of the frank, for example. See
Common Cause, et al. v. William F. Bolger, et al. Civil Action No.
1887-73 in the United States District Court for the District of Columbia. See also the programmatic suggestions and accompanying
discussion in an article by two Common Cause officials, David
Cohen and Wendy Wolff, "Freeing Congress from the Special Interest State: A Public Interest Agenda for the 1980's," *Harvard
Journal on Legislation* 17 (Spring 1980), pp. 253–93.

Chapter V: Political Reform and Democratic Values

1. See Kenneth A. Bode and Carol F. Casey, "Party Reform: Revisionism Revisited," in Robert A. Goldwin (ed.), *Political Parties in
the Eighties* (Washington, D.C.: American Enterprise Institute,
1980), pp. 3–19; Donald M. Fraser, "Democratizing the Democratic Party," *ibid.*, pp. 116–32; William J. Crotty, *Decision for the
Democrats* (Baltimore: Johns Hopkins University Press, 1978); De-
metrios Caraley et al., "American Political Institutions after Watergate—A Discussion," *Political Science Quarterly* 89 (Winter 1974–
75), pp. 713–49; Charles Longley, "National Party Reform and
the Presidential Primaries," prepared for delivery at the 1981 Annual Meeting of the Mid-West Political Science Association, April
15–19, 1981, Cincinnati; Norman C. Miller, "Democratic Reforms: They Work," *Wall Street Journal* (May 16, 1972); David S.
Broder, "From the Grass Roots Up," *Washington Post* (January 23,
1980); Broder, *The Party's Over* (New York: Harper & Row, 1972),
pp. 230–32.
2. V. O. Key, Jr., *American State Politics* (New York: Knopf, 1956), p.
133.
3. *Ibid.* p. 152.
4. See Austin Ranney, *The Federalization of Presidential Primaries*
(Washington, D.C.: American Enterprise Institute, 1978); and
Ranney, *Curing the Mischiefs of Faction* (Berkeley: University of
California Press, 1975); James I. Lengle, *Representation and Presidential Primaries: The Democratic Party in the Post Reform Era* (Westport, Conn.: Greenwood, 1981); Commission on Presidential
Nomination and Party Structure (Morley Winograd, chairman),

Openness, Participation and Party Building: Reforms for a Stronger Democratic Party (Washington D.C.: Democratic National Committee, January 25, 1978).
5. Lengle, *Representation and Presidential Primaries*.
6. Richard F. Schier, "Take Me Out to the Pol Game," *New York Times* (July 17, 1979).
7. *Ibid.*
8. Haynes Johnson, "Portrait of the New Delegate," *Washington Post* (July 8, 1972). Most studies of the "representativeness" of delegates stress delegates' political attitudes rather than their social standing. See, for example, Jeane Kirkpatrick, *The New Presidential Elite* (New York: Russell Sage Foundation, 1976); Mark Stern, Sandra Guest, Roger Handberg, William Maddox, "Party Reform and Party Change," presented at Mid-West Political Science Association Meeting, April 15–19, 1981, Cincinnati; Barbara G. Farah, "Convention Delegates: Party Reform and the Representativeness of Party Elites, 1972–1980," prepared for delivery at the 1981 Annual Meeting of the American Political Science Association, New York, September 3–6, 1981.
9. A good introductory discussion of some of the issues involved in what has become an enormous and very esoteric literature is Dennis C. Mueller, *Public Choice* (Cambridge: Cambridge University Press, 1979), especially pp. 19–67.
10. Perhaps the most famous discussion of this problem in the literature of political science is contained in Robert A. Dahl, *A Preface to Democratic Theory* (Chicago: University of Chicago Press, 1956), pp. 90–119. See also John D. May, "Up with Dick Daley," *Intellectual Digest* (March 1973), pp. 85–87.
11. See Steven J. Brams and Peter C. Fishburn, "Approval Voting," *American Political Science Review* 72 (September 1978), pp. 831–47; Brams, *The Presidential Election Game* (New Haven: Yale University Press, 1978), pp. 193–229; and Brams, "Approval Voting: A Practical Reform for Multicandidate Elections," *National Civic Review* 68 (November 1979), pp. 549–53, 560. As with any automated process, there are problems associated with approval voting in that its proponents are probably wrong in thinking such a system guarantees that centrists or moderates would be favored. Moderates win under approval voting only if voters who favor extremist candidates also cast their votes for moderates. But if centrist voters, being moderate, find a wide range of alternatives acceptable and ideological voters meanwhile bullet-vote, an extremist candidate could end up the winner.

12. See James I. Lengle, "Divisive Presidential Primaries and Party Electoral Prospects, 1932–1976," *American Politics Quarterly* 8 (July 1980), pp. 261–78; Robert A. Bernstein, "Divisive Primaries Do Hurt: U.S. Senate Races, 1956–1972," *American Political Science Review* 71 (June 1977), pp. 540–45; Andrew Hacker, "Does a 'Divisive' Primary Harm a Candidate's Election Chances?" *American Political Science Review* 59 (March 1965), pp. 105–10; Donald Johnson and James Gibson, "The Divisive Primary Revisited: Party Activists in Iowa," *American Political Science Review* 68 (March 1974), pp. 67–77.

13. The most pertinent historical example of this problem at work is no doubt the nomination of Senator George McGovern by the Democrats in 1972. It was not delegates from primaries alone who selected McGovern as the nominee. The primaries did, however, reveal a pattern of preferences within the party that, while selecting McGovern, foreshadowed his landslide defeat. Brokerage mechanisms at the national convention were not available to deal with the impending disaster. See Vermont Royster, "Can McGovern Overcome the Odds?" *Wall Street Journal* (July 13, 1972); Fred L. Zimmerman, "Unhappy Days: A Mood of Bitterness Deepens at Convention as Squabbles Drag On," *Wall Street Journal* (July 11, 1972); Arlen J. Large, "Divided Democrats: Party Likely To Emerge from Its Convention with Severe Wounds," *Wall Street Journal* (July 10, 1972); George Gallup, "No Democrat Shows Ability To Attract Votes Like Nixon," *Washington Post* (July 2, 1972); Jude Wanniski, "The McGovernites and Darwin's Law," *Wall Street Journal* (July 10, 1972); Arlen J. Large, "And Now McGovern Must Seek To Charm Another Constituency," *Wall Street Journal* (June 8, 1972); David S. Broder, "The Splintered Majority," *Washington Post* (July 9, 1972); Frank Lynn, "State Democratic Factions Fail To Unite on McGovern," *New York Times* (October 8, 1972); Ronald Sullivan, "Lag in Student-Vote Drive Laid to McGovern's Staff," *New York Times* (October 2, 1972); Jack Rosenthal, "Yankelovich Says Polls May Create a Bandwagon," *New York Times* (October 6, 1972). At its first opportunity following the 1972 debacle the Democratic party once again tinkered with its rules for delegate selection under the aegis of a new commission headed by Representative Barbara Mikulski. The Mikulski rules abolished winner-take-all primaries and encouraged proportional representation in caucuses and conventions in an attempt to keep in the running strong candidates who did not win early primaries but made a respectable showing. See David S. Broder, "New Democratic Rules: First-Ballot Victory Unlikely," *Washington Post* (March

31, 1974). The effort has proven unsuccessful. Early delegate commitments stimulated by news media coverage of early decisions seem to have prevailed, and the role of the national convention has remained vestigial. Thus the problem of aggregation described in the text remains.

14. The "right winner" in this example would *either* be (1) the winner of all pair-wise comparisons, if such a candidate exists or (2) if there is no such winner, the candidate who emerges after a process in which delegates are allowed to deliberate, change their preferences for strategic purposes, and respond to the views of other delegates.

15. Key, *American State Politics*, pp. 144–45.

16. See, for example, Donald Fraser, "Democratizing the Democratic Party."

17. See Michael J. Robinson, "TV's Newest Program: 'The Presidential Nominations Game,'" *Public Opinion* 1 (May–June 1978), pp. 41–46; Michael J. Robinson and Karen A. McPherson, "Television News Coverage before the 1976 New Hampshire Primary: The Focus of Network Journalism," *Journal of Broadcasting* 21 (Spring 1977), pp. 177–86; Michael J. Robinson and Karen A. McPherson, "The Early Presidential Campaign on Network Television," in A. Casebier and J. J. Casebier (eds.), *Social Responsibilities of the Mass Media* (Washington, D.C.: University Press of America, 1978), pp. 5–41.

18. This was suggested both by the Democratic party's post-1976 Winograd Commission and its post-1980 Hunt Commission. In both cases exceptions were given to the states of Iowa and New Hampshire.

19. As indeed they already do, given the overwhelming influence of early results on later alternatives. *Congressional Quarterly Weekly Report* (December 26, 1981), p. 2567 shows:

Table 5.4 States seek early primaries

	1968		1972		1976		1980	
Democratic Primaries February to early April	2	(12%)	4	(17%)	8	(27%)	13	(37%)
Late April to June 1	10	(59%)	14	(61%)	19	(63%)	14	(40%)
Rest of June	5	(29%)	5	(22%)	3	(10%)	8	(23%)
	17	(100%)	23	(100%)	30	(100%)	35	(100%)

Table 5.4 (*continued*)

	1968		1972		1976		1980	
Republican Primaries								
February to early								
April	2	(13%)	4	(17%)	7	(25%)	11	(34%)
Late April to								
June 1	9	(56%)	13	(57%)	18	(64%)	13	(41%)
Rest of June	5	(31%)	6	(26%)	3	(11%)	8	(25%)
	16	(100%)	23	(100%)	28	(100%)	32	(100%)

Source: *Congressional Quarterly.*

20. This also was a proposal of the Democratic party's Hunt Commission.
21. See Richard L. Rubin, *Press, Party and Presidency* (New York: Norton, 1981), pp. 191–96. "Not only was [television journalists'] affirmation of primaries clear from the vastly disproportionate air time given primaries compared to other selection methods, but also numerous phrases attributing inherent democratic values to primaries appeared, sprinkled liberally throughout network news" (p. 193).
22. *Cousins v. Wigoda* 419 *U.S.* 477 (1975); *Democratic Party of U.S. v. La Follette* 67 *L. ED.2d* 82 (1981).
23. There is a $50,000 expenditure limit on candidates and their immediate families for those presidential candidates accepting public funding according to the 1976 amendments. Thus, according to P.L. 94-283, Section 305: ". . . no candidate shall knowingly make expenditures from his personal funds, or the personal funds of his immediate family, in connection with his campaign for nomination for election to the office of President in excess of, in the agregate, $50,000." See also *Buckley et al. v. Valeo* 424 U.S. 1 (1976).
24. There was evidence, of course, of political contributions by politically interested persons and groups, and of the already illegal collection of funds. Readers will judge for themselves the plausibility of the inferences constructed of these materials by the District of Columbia Circuit Court in *Buckley v. Valeo* 519 F.2nd 821 (D.C. Circuit, 1975) and the Supreme Court in *Buckley et al. v. Valeo*. The Circuit Court opinion gives excellent guidance through the evidence on this point in the legislative record. A brilliant defense of the Circuit Court's position is Harold Leventhal, "Courts and Political Thickets," *Columbia Law Review* 77 (April 1977), pp. 345–87. Leventhal, a former treasurer of the Democratic party, is

thought to have been the judge of the District of Columbia Circuit most influential in writing the per curium—unsigned—opinion of the Circuit Court. A consideration of the constitutional issues raised by *Buckley v. Valeo* less indulgent to the view of the D.C. Circuit is Daniel D. Polsby, *"Buckley v. Valeo:* The Special Nature of Political Speech," *Supreme Court Review 1976* (Chicago: University of Chicago Press, 1977), pp. 1–44.

25. *Democratic Party of U.S. v. La Follette.*
26. See the careful analysis of this and related issues by one of Wisconsin's foremost political scientists, Leon D. Epstein, "Party Confederations and Political Nationalization," prepared for the conference on the Continuing Legacy of the Articles of Confederation, Center for the Study of Federalism, Temple University, Philadelphia, August 30–September 2, 1981.
27. William L. Riordan, *Plunkitt of Tammany Hall* (New York: Dutton, 1963) (first printed 1905), pp. 3–6.
28. *Ibid.*, pp. 11–16 and *passim.*
29. A good brief account of the response of the national convention is Richard H. Rovere, "Letter from Washington," *The New Yorker* (October 16, 1965), pp. 233–44. See also Lester M. Salamon and Stephen Van Evera, "Fear, Apathy, and Discrimination: A Test of Three Explanations of Political Participation," *American Political Science Review* 67 (December 1973), pp. 1288–1306; and Fraser, "Democratizing the Democratic Party."

Index

Executive Office of the President, 221;
in Nixon administration, 94
"Expected," as phantom candidate,
144

F.C.C., 83
Factionalism: in Democratic party,
85-86; in Republican party, 85
Factions, and coalitions, 64-71
Farah, Barbara G., 249
Farney, Dennis, 207, 244
Favorite son candidacies, 15, 56, 71;
and primaries, 57
FBI, 218
Federal Election Campaign Act
(FECA) of 1971, 38
Federal Election Campaign Act
Amendments of 1974, 36-37, 59, 83;
and limitations of financial con-
tributions, 178-79; and public
finance of campaigns, 37; Supreme
Court challenges to, 38
Federal Election Campaign Act
Amendments of 1979, 81
Federal Election Commission (FEC),
39, 61, 74, 81, 202, 205-6, 208, 229;
creation of, 38; Supreme Court
challenge, 38
Federal subsidies: candidates in pri-
maries receiving, 59; and FECA
1974 Amendments, 61
Federalist, The, 95, 203, 221
Fenno, Richard F., Jr., 199, 218, 219,
223
Ferber, Mark F., 228
Fields, Cheryl M., 220
Finch, Robert, 91
Fishburn, Peter C., 249
Fisher, Louis, 221
Fleishman, Joel L., 202, 211
Florida, 12; and Ford campaign, 80;
primary, 1976, 129, 204
Fogarty, John, 236
Ford, Gerald, 11, 79, 115, 134, 206,
229, 235; 1976 campaign, 79, 80, 85
Ford, William, 232-33
Foreign Ministry, West Germany,
231-32
France, Joseph, 11
France, riots and fall of government in
1968, 17

Fraser, Donald M., 195-96, 198, 200-
201, 248, 251, 253
Free Soil party, 209
Fresca, 100
Fried, Edward R., 221
Friendly, Jonathan, 243
Friends of the Earth, 247
Fritchey, Clayton, 231
Fuller, Jack, 238
Fund-raisers, as political decision mak-
ers, 73

Galbraith, John Kenneth, 192, 198
Gallup, George, 216, 250
Gallup Opinion Index: December
1980, 212; November 1981, 213-14;
Report, 1980, 87
Gallup poll: January 1968, 192; Feb-
ruary 1968, 192; February 1968, on
Vietnam, 19; March 1968, 192;
March 1968, on Vietnam, 19-20;
April 1968, 194; April 28, 1968, 23;
May 1968, 194; August 1979, 229
Gallup Report News Release, June
1968, 194
Gans, Herbert J., 241
Gardner, John, 116
Garfield, James A., assassination, 181
Gartner, Michael, 202
Gelb, Leslie H., 223
Georgia: 1972 delegate selection, 56;
reason for primary, 57
Georgians, in Carter cabinet, 103-5
Germond, Jack W., 205, 236
Getler, Michael, 231, 232
Gibson, James, 250
Gilligan, John, 56
Glad, Betty, 225
Goldman, Ralph M., 11, 189, 190
Goldschmidt, Neal, 128
Goldwater, Barry, 11
Goldwin, Robert A., 187, 200, 248
Goodwin, Richard, 25
Gorman, Joseph Bruce, 190, 191
Gottlieb, Stephen E., 211
Grabowicz, Paul, 206
Grass roots activity, 81
Great Society, 105
Green, Bill, 242
Greenback party, 209
Greenberg, Daniel S., 219
Greenfield, Meg, 193, 194, 224-25

260

Index

Greenstein, Fred I., 222-23, 239, 240
Greider, William, 223
Guatemala, U.S. Ambassador to, killed, 17
Guest, Sandra, 249
Guyer, Tennyson, 207

Hacker, Andrew, 250
Hagen, Tim, 232-33
Haggerty, Brian A., 200
Haig, Alexander M., Jr., 245
Halberstam, David, 192, 193, 194, 243, 245
Hamilton, Alexander, 221
Hamlet, 121
Handberg, Roger, 249
Hardin, Clifford, 91
Harding, Warren G., 170, 188
Hardwick, Elizabeth, 197, 198
Harriman, Averell, 10
Harris, Fred, 236
Harris, Louis, 198, 217
Harris, Patricia Roberts, 103, 223-24
Harrison, Gilbert, 193-94
Hart, Gary, 136, 238
Harvard University, 135
Harwood, Richard, 194, 216, 224
Hassan, Crown Prince, 243
Havard, William C., 203
Heard, Alexander, 199
Heclo, Hugh, 247
Heineman, Ben W., Jr., 227
Hennessy, Bernard C., 215
Henry, Laurin L., 218
Herring, E. Pendleton, 222
Hershey, Robert D., Jr., 231-32
Herzog, Arthur, 192
Hess, Stephen, 236, 238-39, 244
Hessler, Curtis A., 227
Hickel, Walter, 91
Hicks, Nancy, 238
Hilton Hotel (Chicago), 197
Hoagland, Jim, 231
Hodgson, Godfrey, 192, 193, 196
Hodgson, James, 92
"Honest graft," 181
Hoover, Herbert, 11, 191
Hornblower, Margot, 223
Horowitz, Irving Louis, 199
House, Karen Elliott, 223
House of Representatives: Committee on Post Office and Civil Service, 221,

Subcommittee on Manpower, 220; Democratic caucus, 149; Ways and Means Committee, 107
Houston, Texas, 205
Hovey, Graham, 224-25
Huckaby, Stan, 206
Hucker, Charles W., 209
Hue, battle of, 17
Hughes, Harold, 26, 195
Hughes, Richard, 32
Hughes Commission, 33, 34, 194, 196, 241; recommendations, and Rules Committee, 28
Humphrey, Hubert, 14, 23, 24, 25, 29, 32, 33, 36, 193, 194, 195, 197, 198; attempts to build bridges during 1968 Convention, 33; delegate advantage, 24; hints at candidacy, 22; and Johnson loyalists, 24; on primaries, 14; resentment toward, 26-27, 32; and Robert Kennedy assassination, 24, 26; on strategy in primaries, 22-24; and support in the South before the convention, 24
Hunt, Albert R., 188, 231, 224-25
Hunt Commission, 177, 251, 252
Huntington, Samuel P., 241
Hussein, King, 243
Hyman, Herbert H., 191, 216

Illinois, 12; delegate selection, 1972, 201; delegation, 1972, 199; primary, 1972, 217; delegate selection, 1976, 204
Impounded funds, under the Nixon administration, 93
Indiana, 1968: and irregular district conventions, 27; and Robert Kennedy, 21
Intensity problem, and primary elections, 164
Interest groups: Carter's relations with, 125-28; media approval of, 133
Iowa: caucus, and news media coverage, 68-69; caucuses, 1976, 129, 236, 251; state convention, 1972, 60; state convention, 1976, and uncommitted delegates, 60
Iran, 115; hostage crisis, 130
Iraq, coup in 1968, 16
Ireland (Northern), riots in 1968, 16
Italy: Chamber of Deputies, 106; general strike in 1968; 16

North Vietnamese army in South Vietnam in 1968, 17
Novak, Robert, 205, 219, 221, 244

Obey, David, 232-33
O'Brien, Lawrence, 25, 198, 200-201, 222
Odegard, Peter H., 240
Office of Economic Opportunity, in Nixon administration, 93
Office of Management and Budget, 206; in Nixon administration, 92-93
Ohio, 84; 1972: delegate selection, 56, McGovern campaign, 68; election law, 56
Oklahoma: 1968 irregular county conventions in, 27; 1972 McGovern delegates from, 68
O'Neill, Thomas P., Jr., 109, 113, 227; on Hamilton Jordan, 126
OPEC, 115
Oppenheimer, Bruce I., 214, 226, 228
Oregon, 1948 primary, 9, 14
Ornstein, Norman J., 215, 226, 228, 247
Orren, Gary R., 200
Osborne, John, 235
Oslo, 231-32
Otten, Alan L., 29, 189, 196, 223
Overacker, Louise, 190
Oxley, Michael, 207

Page, Bruce, 192, 193, 196
Pakistan, general strike in 1968, 16
Panama, coup in 1968, 16
Papan, Louis, 247-48
Paris, university closed in 1968, 17
Participatory Convention selection system, defined, 35
Party caucus selection system, defined, 34
Party leaders: Democratic party, 14; displacement of, 73-75; and John F. Kennedy, 15; and presidential nominee, 14
Patterson, James T., 228
Patterson, Thomas E., 143, 242
Peabody, Robert L., 226
"Peanut brigade," Jimmy Carter's, 104
Peer review, and presidential selection, 169-71
Pennsylvania: delegate primary, 35; McGovern 1972 campaign, 68

Perry, James M., 231, 235
Personal candidate organizations: strategies in building, 67-68; versus party organizations, 72
Peru, coup in 1968, 16
Peterson, Bill, 229
Petrocik, John R., 240-41
Pfeiffer, Jane Cahill, 223
Philadelphia Phillies, 190
Phillips, Cabell, 218
Pierpoint, Robert, 243-44
Pierson, John, 222
Plains, Georgia, 227
Plissner, Martin, 213, 215
Plunkitt, George Washington, 181
Poland, riots in 1968, 16
Political Action Committees (PACs), 74
Political decision-makers, new, 73-75
Political intermediaries: erosion of traditional, 141-42; purposes and functions, 138
Political intermediation, reform and changes in, 139-40
Poll analysts, as new political decision-makers, 73
Polsby, Daniel D., 199, 211, 252-53
Polsby, Nelson W., 91, 114, 187, 188, 191, 199, 216, 217, 221, 226, 228, 234, 235-36, 239, 240, 245
Pomper, Gerald M., 200, 213, 215
Pool, Ithiel de Sola, 191, 240
Popkin, Samuel, 191
Populist party, 209
Powell, Jody, 109, 120, 121
Prague, 32
Presidential ambassadors, as cabinet members, 99-101
Pride, Richard A., 197
Primaries, 14; in 1952, 11, 13-14; advantages of early entry, 60; as appeals against leaders, 15; pre-1968, 16; proliferation of, 56-59; and weakening grasp of parties, 66
Progressive party, 209
Prohibition party, 209
Public campaign financing: and FECA of 1971, 38; and FECA 1974 amendments, 37; and third parties in the U.S., 83
Public opinion surveys, and campaign management, 74
Public relations specialists, as political decision makers, 73